T0201179

# PROBABILISTIC
# RELIABILITY MODELS

# PROBABILISTIC RELIABILITY MODELS

IGOR USHAKOV

A JOHN WILEY & SONS, INC., PUBLICATION

*Library of Congress Cataloging-in-Publication Data:*

Ushakov, I. A. (Igor Alekseevich)
    Probabilistic reliability models / Igor A. Ushakov.
        p. cm.
      Includes bibliographical references and index.
    ISBN 978-1-118-34183-4 (hardback)
    1. Reliability (Engineering)–Mathematical models. 2. Probabilities. I. Title.
TA169.U84 2012
320'.00452015192—dc23

                                                                            2012005946

Printed in the United States of America

10 9 8 7 6 5 4 3 2 1

*To Boris Gnedenko, a mentor throughout my life*

# CONTENTS

# PREFACE

I dedicate this book to a great man who was more than my mentor. He took the place of my father who passed away relatively early. Boris Gnedenko was an outstanding mathematician and exceptional teacher. In addition, he was a magnetic personality who gathered around him tens of disciples and students. He was a founder of the famous Moscow Reliability School that produced a number of first-class mathematicians (among them Yuri Belyaev, Igor Kovalenko, Jacob Shor, and Alexander Solovyev) and talented reliability engineers (including Ernest Dzirkal, Vadim Gadasin, Boris Kozlov, Igor Pavlov, Allan Perrote, and Anatoly Raykin).

Why did I write this book?

Since the beginning of my career, I have been working at the junction of engineering and mathematics—I was a reliability engineer. As an engineer by education, I never had a proper mathematical background; however, life has forced me to submerge in depth into the area of probability theory and mathematical statistics. And I was lucky to meet at the start of my career the "three pillars on which rested the reliability theory" in Russia, namely, Boris Gnedenko, Alexander

Solovyev, and Yuri Belyaev. They helped me understand the nuances and physical sense of many mathematical methods.

Thus, I have decided to share with the readers my experience, as well as many real mathematical insights, that happened when I submerged myself into reliability theory.

Boris Gnedenko once told me: "Mathematical reliability models are engendered by practice, so they have to be adequate to reality and should not be too complex by their nature."

To get an understanding of "real reliability," one goes through a series of painful mistakes in solving real problems. Engineering intuition arrives to mathematicians only after years of working in reliability engineering. At the same time, proper mathematical knowledge comes to reliability engineers after multiple practical uses of mathematical methods and having experienced "finger sensation" of formulas and numbers.

I remember my own thorny way in the course of my professional career. In writing this reliability textbook, I have tried to include as much as possible "physical" explanations of mathematical methods applied in solving reliability problems, as well as "physical" explanations of engineering objects laid on the basis of mathematical models.

At the end of the book, the reader can find a wide list of monographs on reliability. I must, however, note a few books that, in my opinion, are basic in this area. They are (in order of publication) the monographs by Igor Bazovsky (1961), David K. Lloyd and Myron Lipow (1962), Richard Barlow and Frank Proschan (1965), and Boris Gnedenko, Yuri Belyaev, and Alexander Solovyev (1965). These books cover the entire area of probabilistic reliability modeling and contain many important theoretical and practical concepts.

IGOR USHAKOV

*San Diego, California*
*March 31, 2012*

# ACRONYMS AND NOTATION

## ACRONYMS

| | |
|---|---|
| AC | Availability coefficient |
| d.f. | Distribution function |
| DFR | Decreasing failure rate |
| FR | Failure rate |
| GF | Generating function |
| i.i.d. | Independent and identically distributed (about r.v.) |
| IFR | Increasing failure rate |
| LT | Laplace transform |
| MDT | Mean downtime |
| MTBF | Mean time between failures |
| MTTF | Mean time to failure |
| OAC | Operational availability coefficient |
| PEI | Performance effectiveness index |
| PFFO | Probability of failure-free operation |
| r.v. | Random variable |
| RBD | Reliability block diagram |
| TTF | Random time to failure |
| UGF | Universal generating function |

## NOTATION

| | |
|---|---|
| $F(t)$ | Distribution function |
| $K$ | System stationary availability coefficient |
| $K(t)$ | System nonstationary availability coefficient |
| $p_k(t)$ | Probability of failure-free operation of unit $k$ |
| $P(t)$ | Probability of system's failure-free operation |
| $q_k(t)$ | Probability of failure of unit $k$ |
| $Q(t)$ | Probability of system's failure |
| $T$ | Mean time to/between failures |
| $X$ | Random variable |
| $\lambda$ | Failure rate |
| $\tau$ | Downtime |
| $\xi$ | Random time to/between failures |
| $\eta$ | Random recovery time |
| $\psi(\cdot)$ | System's structural function |
| $\displaystyle\sum_{1\le k\le n}$ | Sum by $k$ from 1 to $n$ |
| $\displaystyle\prod_{1\le k\le n}$ | Product by $k$ from 1 to $n$ |
| $\cup$ | Logic sum ("or") |
| $\cap$ | Logic product ("and") |
| $\displaystyle\bigcup_{1\le k\le n}$ | Logic sum by $k$ from 1 to $n$ |
| $\displaystyle\bigcap_{1\le k\le n}$ | Logic product by $k$ from 1 to $n$ |
| $i = 1, \ldots, n$ | Set of natural numbers from 1 to $n$ |

# 1

# WHAT IS RELIABILITY?

## 1.1 RELIABILITY AS A PROPERTY OF TECHNICAL OBJECTS

Reliability of a technical object is its ability to perform required operations successfully. Usually, it is assumed that an object is used in accordance with its technical requirements and is supported by appropriate maintenance.

One of the outstanding Russian specialists in cybernetics, academician Axel Berg, has said: "Reliability is quality expanded in time."

Reliability is a broad concept. Of course, its main characterization is the failure-free operation while performing required tasks. However, it also includes such features as availability, longevity, recoverability, safety, survivability, and other important properties of technical objects.

Speaking of reliability, one has to introduce a concept of failure. What does it mean—"successful operation?" Where is the limit of "successfulness?"

*Probabilistic Reliability Models*, First Edition. Igor Ushakov.
© 2012 John Wiley & Sons, Inc. Published 2012 by John Wiley & Sons, Inc.

In reliability theory, usually one analyzes systems consisting of units, each of which has two states: operational and failure. If some "critical" set of units has failed, it leads to system failure. However, a unit's failure does not always lead to "total" system failure; it can decrease its ability, but main system parameters still could be in appropriate limits.

However, such "instantaneous" failure is only one of the possibilities. The system can fail due to monotonous drifting of some parameters that can bring the entire system to the unacceptable level of performance.

In both cases, one needs to formulate failure criteria.

## 1.2  OTHER "ILITIES"

Reliability itself is not the final target of engineering design. An object can be almost absolutely reliable under "greenhouse conditions"; however, at the same time, it can be too sensitive to real environment. Another situation: an object is sufficiently reliable but during operation it produces unacceptable pollution that contaminates natural environment.

Below we discuss some properties closely connected to the concept of reliability.

- *Maintainability.* Failure-free operation is undoubtedly a very important property. However, assume that a satisfactorily reliable object needs long and expensive restoration after a failure. In other words, maintainability is another important property of recoverable systems. Maintainability, in turn, depends on multiple factors.

    The quality of restoration of an object after failure as well as time spent on restoration significantly depends on repairmen qualification, availability of necessary tools and materials, and so on.

- *Safety.* Development of large-scale industrial objects attracts attention to safety problem. It is clear that not only an object has to perform its main operating functions, but it is also very important that the "successful operation" is not dangerous for personnel's health and does not harm ecology.

One of the most tragic events of this kind occurred in 1984. It was the Bhopal Gas Tragedy—one of the world's worst industrial catastrophes. It occurred at the Union Carbide India Limited pesticide plant in India. The catastrophe led to almost immediate death of about 7000 people and about 8000 died from gas-related diseases. In addition, over half a million people got serious injuries.

Then, in 1986 explosion and fire occurred at the Chernobyl Nuclear Power Plant in the former Soviet Union. Large quantities of radioactive contamination were released into the atmosphere, which spread over much of Western USSR and Europe. It is considered the worst nuclear power plant accident in history. Thousands of workers were killed almost instantaneously, and about 1 million cancer deaths occurred between 1986 and 2004 as a result of radioactive contamination.

Actually, problem of safety appears not only in the context of failures. A number of "reliable" industrial plants are extremely unsafe for the people who work there or live in the area (Figure 1.1).

- *Survivability.* The problem of survivability is very close to the reliability and safety problems. This is an object's property to survive under extreme natural impacts or intentional hostile actions.

  In this case, nobody knows the moment of disaster, so an object has to have some "warranty level" of safety factor. In our time, the survivability problem is extremely important for large-scale terrestrial energy systems.

  The 1999 Southern Brazil blackout was the largest power outage ever. The blackout involved Sao Paulo, Rio de Janeiro, and other large Brazilian cities, affecting about 100 million people.

  Then in 2003 there was a widespread power outage known as the Northeast blackout. It was the second most widespread blackout in history that affected 50 million people in Canada and the United States.

**FIGURE 1.1**    Typical "industrial landscape" with terrible air pollution.

On March 11, 2011, a ferocious tsunami spawned by one of the largest *earthquakes* ever recorded slammed Japan's eastern coast. This earthquake, officially named the Great East Japan Earthquake, was 9 magnitudes on the Richter scale. Tsunami waves reached up to 40 meters, struck the country, and, in some cases, traveled up to 10 kilometers inland in Japan. States of emergency were declared for five nuclear reactors at two power plants. There were some severe damages, although consequences were much less than those after Chernobyl.

Problem of survivability has become essential in our days when unpredictable by location and strength terrorist acts are initiated by religious fanatics.

- *Stability.* An object performs under unstable conditions: environment can change, some simultaneously performing operations can conflict with each other, some disturbances can occur, and so on. An object has to have an ability to return to normal operational state after such inner or outer influences.

- *Durability.* Reliability as a concept includes such a property as durability. For instance, mechanical systems, having some fractioning parts, can be very reliable during the first several hundred hours; however, after some period of time due to wearing out processes they fail more and more frequently, and became unacceptable for further use.

- *Conservability.* This is the property of the object to continuously maintain the required operational performance during (and after) the period of storage and transportation. This property is important for objects that are kept as spares or are subjects of long transportation to the location of the use.

## 1.3   HIERARCHICAL LEVELS OF ANALYZED OBJECTS

Analyzing reliability, it is reasonable to introduce several hierarchical levels of technical objects. Below we will consider systems, subsystems, and units. All these terms are obvious and understandable; nevertheless, we will give some formal definitions for further convenience.

A *unit* is an indivisible ("atomic") object of the lowest hierarchical level in the frame of current reliability analysis.

A *system* is an object of the highest hierarchical level destined for performing required tasks.

Of course, concepts of unit and system are relative: a system in one type of analysis can be a unit in consideration of a large-scale object, and vice versa. In addition, sometimes it is reasonable to introduce an intermediate substance—subsystem. It can be a part of a system that is destined for performing a specific function or a separate constructive part.

System reliability indices can be expressed through corresponding indices of its units and subsystems.

## 1.4   HOW CAN RELIABILITY BE MEASURED?

Reliability can be and has to be measured. However, what measures should be used for reliability?

Distance can be measured in kilometers and miles, weight in kilograms and pounds, and volume in liters and gallons. What kinds of index or indices are appropriate for reliability?

Of course, reliability index depends on the type of a technical object, its predestination, and regime of operating, as well as on some other factors that are usually rather individual.

Generally speaking, all technical objects can be divided into two main classes: unrecoverable and recoverable. All single-use technical objects are unrecoverable. For instance, anti-aircraft missile is used only once. It can be characterized by the probability that the required operation is completed.

A reconnaissance satellite is also a single-use object. However, for this object the best reliability index is an average time of operating without failure: the more time the satellite is in the orbit, the more useful information will be collected.

Most of technical objects we are dealing with are recoverable ones: they can be restored after a failure and can continue their operations.

Let us consider a passenger jet. It is almost obvious that the most important reliability index is the probability that a jet successfully completes its flight. Of course, one should think about longevity and convenience of technical maintenance, although these indices are undoubtedly secondary.

Let us note that the same object may be considered as recoverable or not depending on the concrete situation. It is clear that for the same passenger jet some critical failure, having been occurred during the flight (for instance, engine failure), cannot be corrected. Thus, in this case a jet should be considered as unrecoverable during a flight.

Anti-missile defense systems work in regime "on duty"; that is, they have to be in an operational state at any arbitrary chosen moment of time. For an airport dispatcher system, it is very important to be in an operational state at some required moment of time and successfully operate during an airplane landing. Thus, for such systems the most important property is availability.

For a passenger bus, probably one of the main reliability characterizations is the duration of failure-free operation because it means that the number of unexpected stops due to failures is minimal. Same reliability index is convenient for trucks: it delivers the best economical efficiency during operations.

For most home appliances, cars, and technical equipments, durability is very important because it saves money of the user. At the same time, one does not need "immortal" personal computer because in 2–3 years it will be anyway obsolete and should be replaced by a modern one. There are several commonsense rules that one should keep in mind while choosing reliability indices:

1. They have to reflect specificity of the object and its operating process.
2. They have to be simple enough and should have an understandable physical sense.
3. They have to be calculable analytically or numerically.
4. They have to be empirically confirmed by special tests or during real exploitation.

The number of indices chosen for characterization of reliability of a technical object should be as limited as possible, since multiple indices can only lead to confusion. Do not use "weighted" indices because they usually have no physical sense.

## 1.5  SOFTWARE RELIABILITY

Software reliability is a special topic. Frankly speaking, there is too much confusion and misunderstanding.

Nobody doubts that reliability in technical context is a concept associated with time and randomness. If there is an object (especially, immaterial) that exists in sense "beyond the time" and its failure does not occur randomly, how we can talk about reliability?

Take a look: what is software? It is a set of commands arranged in a special order. It reminds of a book "written" for a "hardware reader" that can "read" it when and if needed.

Is it possible to say about "reliability of a book," keeping in mind its contents? Of course, a book can contain errors ("failures") but these errors are everlasting property of this specific book! These errors can be deleted in the next edition of the book but they are and they will remain forever in this particular edition.

The same picture we observe with software if some "inner programs conflict" or "inconvenient" set of input data appears again and again, which will lead to repeating failures. And it does not depend on current time, and it is not random at all.

For software, we should say about quality, which depends on programmer's qualification and carefulness of testing. To say about "frequency of software failures" is hardly correct.

### 1.5.1   Case Study: Avalanche of Software Failures

In 1970s, the author, being an engineer at R&D Institute of the former Soviet Union, participated in the design of an automatic control system for missile defense. Reliability requirements for the system were extremely high.

After design completion, long and scrupulous tests began. Hardware and software were multiply checked and rechecked: the system seemed "absolutely reliable." But all of a sudden, a long series of software failures occurred in a row!

Acceptance Commission was in panic . . .

After careful analysis, it was found that a young lieutenant who was working as an operator mentioned that some sequence of specific commands led to computer fault. He made a corresponding note in a Test Protocol though, being too much curious, continued to try the same commands multiply.

Definitely, recording several tens of faults was unreasonable. Only one fault of software was recorded. Afterward, the software had been corrected . . .

However, there is a question: how you should characterize software reliability? The only fault has been recorded during 50 hours of testing. May you say that the software failure occurs once in 50 hours on average? Moreover, the program had been "repaired." So, does it mean that after this the software became "absolutely reliable"?

Who knows when and how next time such "inconvenient" circumstances may occur in real operating regime?

# 2

# UNRECOVERABLE OBJECTS

## 2.1 UNIT

An indivisible ("atomic") part of a system under current consideration is called a unit. Process of an unrecoverable unit operation is defined by its random time to failure (TTF). Let us denote this random variable (r.v.) by $\xi$ and its distribution function (d.f.) by $F(t) = \Pr\{\xi < t\}$.

### 2.1.1 Probability of Failure-Free Operation

The probability of failure-free operation (PFFO) of an unrecoverable unit during time $t$ is, by definition,

$$P(t) = \Pr\{\xi \geq t\} = 1 - F(t). \tag{2.1}$$

If d.f. is exponential, then

$$P(t) = \exp(-\lambda t). \tag{2.2}$$

*Probabilistic Reliability Models*, First Edition. Igor Ushakov.
© 2012 John Wiley & Sons, Inc. Published 2012 by John Wiley & Sons, Inc.

For a highly reliable unit when $\lambda t \ll 1$, there is a good approximation:

$$P(t) \approx 1 - \lambda t. \tag{2.3}$$

(This estimation gives an understated estimate with error of order $(\lambda t)^2$.)

Sometimes it is reasonable to introduce the so-called "indicator function" that is defined as follows:

$$x(t) = \begin{cases} 1, & \text{if unit is in operational state at moment } t, \\ 0, & \text{otherwise.} \end{cases} \tag{2.4}$$

It is clear that $x$ is a binary Boolean[1] r.v. The unit's PFFO can be defined in new terms as

$$P(t) = E\{x(t)\} = 1 \cdot P(t) + 0 \cdot [1 - P(t)], \tag{2.5}$$

where $E\{\cdot\}$ is an operator of mathematical expectation.

This form of presentation will be sometimes useful below. (For compactness of formulas, we will sometimes omit argument $t$.)

### 2.1.2  Mean Time to Failure

The MTTF of an unrecoverable unit in general case is calculated as

$$T = E\{\xi\} = \int_0^\infty t \cdot f(t) \mathrm{d}t = \int_0^\infty P(t) \mathrm{d}t. \tag{2.6}$$

For exponential d.f., this integral gives

$$T = \int e^{-\lambda t} \, \mathrm{d}t = \frac{1}{\lambda}. \tag{2.7}$$

---

[1] This type of variable is named after George Boole (1815–1864), an English mathematician and philosopher. He invented Boolean logic—the basis of computer logic.

## 2.2  SERIES SYSTEMS

A series system is such a system for which failure of any of its units leads to inevitable failure of the entire system. Usually, these systems present a series connection of their subsystems or units.

The series structure is one of the most common structures considered in engineering practice. In reliability engineering, for describing the logical connection of system units, one uses the so-called reliability block diagrams (RBDs). For a system of $n$ units, RBD can be presented in the form shown in Figure 2.1.

### 2.2.1  Probability of Failure-Free Operation

Denote the system random TTF by $\xi$ and units' TTFs by $\xi_k$, where $1 \leq k \leq n$; then from the definition of a series system follows

$$\xi = \min_{1 \leq k \leq n} \xi_k. \tag{2.8}$$

This statement is easily understood from Figure 2.2.

**FIGURE 2.1**   Reliability block diagram for a series system.

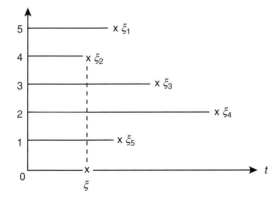

**FIGURE 2.2**   Illustration of a series system TTF.

Often one uses Boolean equations for description of reliability structures. For a series system, the Boolean expression is

$$\phi(X) = \bigcap_{i=1}^{n} x_i,$$  (2.9)

where $X = (x_1, x_2, \ldots, x_n)$. For independent units, $P(t) = E\{\phi(X(t))\}$, and by the theorem of multiplications, we can immediately write

$$P(t) = \Pr\left\{\bigcap_{i=1}^{n} x_i = 1\right\} = \Pr\{\xi_1 \geq t\} \cdot \Pr\{\xi_2 \geq t\} \cdots \Pr\{\xi_n \geq t\}$$
$$= \prod_{k=1}^{n} \Pr\{\xi_k \geq t\} = \prod_{k=1}^{n} p_k(t),$$

(2.10)

where the probability of failure-free operation of unit $k$ is denoted by $p_k(t) = \Pr\{\xi_k \geq t\}$.

Let us introduce notation $q_k(t) = 1 - p_k(t)$. If system's units are highly reliable, that is, $\max_{1 \leq k \leq n} q_k(t) \ll 1/n$, then

$$P(t) = \prod_{k=1}^{n} [1 - q_k(t)] \approx 1 - \sum_{i=1}^{n} q_k(t).$$  (2.11)

From formula (2.10), one can make the following conclusions:

- A series system's reliability decreases (increases) if the reliability of any unit decreases (increases).
- A series system's reliability decreases (increases) if the number of units decreases (increases).
- A series system's reliability is worse than the reliability of any of its units.

If each unit has exponential d.f. of TTF, then for a series system consisting of such units, one can write

$$P(t) = \prod_{k=1}^{n} \exp(-\lambda_k t) = \exp\left(-t \sum_{k=1}^{n} \lambda_k\right).$$  (2.12)

**TABLE 2.1   System Reliability Dependence on the System Scale**

| $n$ | 10 | 100 | 1000 | 10,000 |
|---|---|---|---|---|
| $P(t)$ | 0.99005 | 0.904837 | 0.367879 | Practically zero |

For systems, consisting of highly reliable units, for which $t \cdot \max_{1 \le k \le n} \lambda_k \ll 1/n$, one can write a convenient approximation:

$$P(t) \approx 1 - t \sum_{k=1}^{n} \lambda_k. \tag{2.13}$$

If all system units were identical, then

$$P(t) = \exp(-\lambda n t). \tag{2.14}$$

For "feeling" the numbers, consider a system consisting of units with $p(t_0) = 0.999$. In Table 2.1, one can see how reliability of the system decreases with the increase in the number of units.

By the way, from this table, one can see that the approximation formula is practically acceptable for values of $\lambda n t$ of order 0.1.

### 2.2.2   Mean Time to Failure

Now consider the MTTF of a series system. For any unrecoverable series system, the random TTF, $\xi$, can be expressed through units' random TTFs ($\xi_k$) in the following way:

$$\xi = \min_{1 \le k \le n} \xi_k. \tag{2.15}$$

In general case, the MTTF can be found only in a standard way as

$$T = E\{\xi\} = \int_0^\infty p(t) \mathrm{d}t. \tag{2.16}$$

For exponential distributions, there is a closed expression:

$$T = \int_0^\infty \exp\left(-t \sum_{k=1}^{n} \lambda_k\right) = \frac{1}{\sum\limits_{k=1}^{n} \lambda_k}, \tag{2.17}$$

where $\lambda_k$ is the parameter of the corresponding d.f.

For a system with identical units with MTTF equal to $T^*$ for all $k = 1, 2, \ldots, n$, one has

$$T = \left(\frac{n}{T^*}\right)^{-1} = \frac{T^*}{n},\tag{2.18}$$

that is, the system MTTF is inversely proportional to the number of units.

## 2.3  PARALLEL SYSTEM

Another principal structure in reliability theory is a *parallel system*. This system is in operational state until at least one of its units is operational. This type of redundancy is called *loaded redundancy* or even in engineering jargon "hot redundancy." Usually, in practice the operating and all redundant units are assumed to be identical. In addition, all units are assumed independent.

RBD for a parallel system of $n$ units is presented in Figure 2.3.

### 2.3.1  Probability of Failure-Free Operation

Let us denote again the system random TTF by $\xi$ and units' TTFs by $\xi_k$, where $1 \leq k \leq n$; then from the definition of a parallel system follows

$$\xi = \max_{1 \leq k \leq n} \xi_k.\tag{2.19}$$

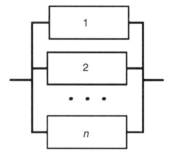

**FIGURE 2.3**    Reliability block diagram for a parallel system.

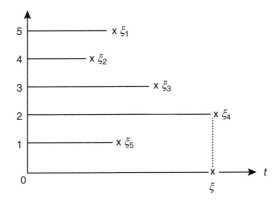

**FIGURE 2.4**   Illustration of a parallel system TTF.

This statement is easy to understand from Figure 2.4.

From the definition of a parallel system follows that it can be described by a Boolean function of the form

$$\phi(\boldsymbol{x}) = \bigcup_{k=1}^{n} x_k = x_1 \cup x_2 \cup \cdots \cup x_n. \tag{2.20}$$

First, we transform expression (2.20) using De Morgan's[2] law of algebra of logics. This law states that

$$x_1 \cup x_2 = \overline{\overline{x}_1 \cap \overline{x}_2}, \tag{2.21}$$

where $\bar{x}$ denotes a complement of $x$. Actually, De Morgan's law can be easily proved by the ancient Greek rule "Look at the drawing." Indeed, look at the diagrams in Figure 2.5 that are called Venn[3] diagrams.

You can see that the second and sixth pictures present the same sets. This law can be easily expanded on arbitrary number of $x$'s. Let us demonstrate it by adding $x_3$:

$$x_3 \cup \overline{\overline{x}_1 \cap \overline{x}_2} = \overline{\overline{x}_3 \cap \overline{\overline{x}_1 \cap \overline{x}_2}} = \overline{\overline{x}_3 \cap \overline{x}_1 \cap \overline{x}_2}, \tag{2.22}$$

[2] Augustus De Morgan (1806–1871) was a British mathematician and logician.
[3] John Venn (1834–1923) was a British logician and philosopher.

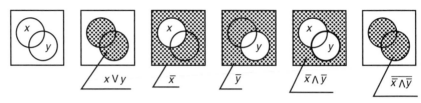

**FIGURE 2.5** Venn diagrams for proving De Morgan's law.

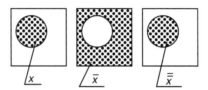

**FIGURE 2.6** Venn diagrams for proving De Morgan's rule of double complement.

where we additionally use another De Morgan rule that double complement of $x$ is $x$ itself. The last statement again is clear from Figure 2.6.

Thus, in general case, one has the following expression:

$$\bigcup_{k=1}^{n} x_k = \overline{\bigcap_{k=1}^{n} \bar{x}_k}. \tag{2.23}$$

From (2.23) follows formula for PFFO:

$$P = E\{\phi(X)\} = 1 - \prod_{i=1}^{n} q_i, \tag{2.24}$$

where $\Pr\{x_k = 0\} = q_k$. The same result follows from the definition of a parallel system: using the theorem of multiplications, one gets

$$Q(t) = \Pr\{(\xi_1 < t) \cap (\xi_2 < t) \cap \cdots \cap (\xi_n < t)\}$$
$$= \prod_{k=1}^{n} \Pr\{\xi_k < t\} = \prod_{k=1}^{n} q_k(t), \tag{2.25}$$

where $Q(t)$ is the probability of parallel system failure, $Q(t) = 1 - P(t)$, and $q_k(t)$ is the probability of unit $k$ failure, $q_k(t) = 1 - p_k(t)$.

Thus, the PFFO of a parallel system is

$$P(t) = 1 - Q(t) = 1 - \prod_{k=1}^{n} q_k(t). \qquad (2.26)$$

Sometimes a different form equivalent to (2.26) is used:

$$\begin{aligned}
P(t) &= p_1(t) + q_1(t) \cdot p_2(t) + q_1(t) \cdot q_2(t) \cdot p_3(t) \\
&\quad + \cdots + q_1(t) \cdot q_2(t) \cdots q_{m-1}(t) \cdot p_m(t) \\
&= p_1(t) + q_1(t) \cdot [p_2(t) + q_2(t) \cdot [p_3(t) + \cdots + q_{m-1}(t) \cdot p_m(t)].
\end{aligned} \qquad (2.27)$$

This expression can be explained as follows:

Pr{a parallel system successfully operates} =
  Pr{the first unit is up during time $t$;
  or
  if the first unit has failed, the second is up during time $t$;
  or
  if both of these units have failed, then the third one;
  and so on}.

From formula (2.26), one can make the following conclusions:

- A parallel system's reliability decreases (increases) if the reliability of any unit decreases (increases).
- A parallel system's reliability decreases (increases) if the number of units decreases (increases).
- A parallel system's reliability is higher than the reliability of any of its units.

If each of the system's unit has an exponential TTF distribution $p_k(t) = \exp(-\lambda_k t)$, for a highly reliable system where $\max\limits_{1 \leq k \leq n} q_k(t) \ll 1/m \ll 1/n$, one can write $q_k(t) \approx \lambda_k t_0$, and, finally,

$$P(t) = 1 - \prod_{k=1}^{n} [1 - \exp(-\lambda_k t)] \approx 1 - t^n \prod_{k=2}^{n} \lambda_k. \qquad (2.28)$$

**TABLE 2.2    Dependence of Parallel System's PFFO on the Number of Parallel Units**

| $n$ | 2 | 3 | 4 | 5 | $\ldots$ | 10 |
|---|---|---|---|---|---|---|
| $P = 0.9$ | 0.991 | 0.9991 | 0.99992 | 0.999992 | $\ldots$ | Practically 1 |

If all units of a parallel system are identical with exponentially distributed TTF, then (2.28) turns to the following:

$$P(t) \approx 1 - (\lambda t)^n. \qquad (2.29)$$

In other words, the distribution function of parallel system TTF has a Weibull[4]–Gnedenko[5] distribution with parameters $\alpha = \lambda^n$ and $\beta = n$ (see Appendix A.2.8).

For "having a sensation" of the numbers, consider a parallel system of $n$ identical units with $p = 0.9$. In Table 2.2, one can see how reliability of the system significantly increases with the increase in the number of units.

### 2.3.2    Mean Time to Failure

The MTTF of a parallel system in general case can be calculated only by integration of corresponding function $P(t)$. However, when each unit has exponential distribution of TTF, an analytical expression can be derived. For this purpose, write the PFFO in the form (2.27). Simple integration immediately gives us the result:

$$T = \sum_{1 \leq k \leq n} \frac{1}{\lambda_k} - \sum_{1 \leq k < j \leq n} \frac{1}{\lambda_k + \lambda_j} + \cdots + (-1)^n \frac{1}{\sum_{1 \leq k \leq n} \lambda_k}. \qquad (2.30)$$

---

[4] Ernst Hjalmar Waloddi Weibull (1887–1979) was a Swedish engineer, scientist, and mathematician. In the middle of 1930s, he suggested a model of "weakest link" type. He introduced a two-parameter distribution of rather universal kind.

[5] Boris Vladimirovich Gnedenko (1912–1995) was an outstanding Soviet mathematician who proved in the very beginning of the 1940s a cycle of limit theorems concerning extreme r.v.'s. The so-called Weibull distribution was a particular case of the entire class of limit distributions. This fact gives rise to call this distribution by two names.

If, in addition, all units are identical, then (2.30) turns into the following:

$$T = \frac{1}{\lambda}\left(1 + \frac{1}{2} + \frac{1}{3} + \cdots + \frac{1}{n}\right) = \frac{1}{\lambda}\sum_{1 \le k \le n}\frac{1}{k}, \qquad (2.31)$$

where $1/\lambda$ is the MTTF of a single unit.

Explanation of this formula is understandable on the basis of the following simple and "physical" arguments. Consider a system consisting of $n$ identical and independent units, each of which has exponential distribution with parameter $\lambda$. Assume that we have a series system of $n$ units but after first failure the system instantly transforms into a system of $(n-1)$ units and continues working with no failure. MTTF of that system is equal to $1/n\lambda$. Next system works on average $1/(n-1)\lambda$ until its failure and instantly transforms into a series system of $(n-2)$ units. Such transformation continues until the last survivor that is working on average time $1/\lambda$.

This procedure is illustrated for a parallel system with $n = 5$ in Figure 2.7.

With unlimited $n$ increase, MTTF of a parallel system approaches infinity, although this increase is very slow. Some numbers are given in Table 2.3.

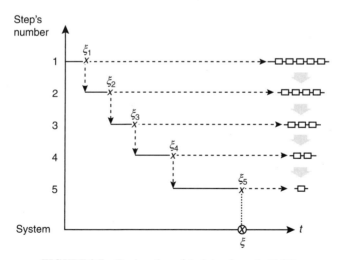

**FIGURE 2.7**   Explanation of deriving formula (2.35).

**TABLE 2.3    Increase of Parallel System MTBF Depending on the Number of Parallel Units**

| Number of Parallel Units | MTTF Increase |
| --- | --- |
| 2 | =1.50 |
| 3 | ≈1.83 |
| 5 | ≈2.28 |
| 10 | ≈2.93 |
| 15 | ≈3.32 |
| 20 | ≈3.60 |

For large $n$, one can use the Euler[6] formula for a harmonic set:

$$\sum_{k=1}^{n} \frac{1}{k} \approx \ln n + C, \tag{2.32}$$

where $C$ is the Euler constant ($C \approx 0.5772$). However, hardly anybody will use multiple loaded redundancy for MTTF increase because it is too ineffective.

## 2.4    STRUCTURE OF TYPE "K-OUT-OF-N"

A system with such a structure consists of $n$ units and remains in operational state until $(n-k+1)$ units have failed. Structural function of such a system can be written as follows:

$$\psi(\mathbf{x}) \begin{cases} 1, & \text{if } \sum_{i=1}^{n} x_i \geq k, \\ 0, & \text{otherwise.} \end{cases} \tag{2.33}$$

Factually, such a system can be considered as a series system of $k$ units with $(n-k)$ redundant units, each of which can replace any one

---

[6] Leonard Euler (1707–1783) was a Swiss, German, and Russian scientist who made significant contributions in mathematics, physics, mechanics, and astronomy. During the second half of his life, he worked at the Saint Petersburg Academy of Sciences, Russia.

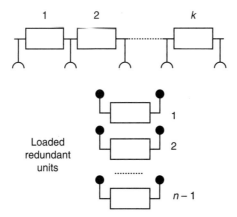

**FIGURE 2.8**  Conditional RBD of an unrecoverable loaded "k-out-of-n" system.

of the failed operating units. Conditional RBD of this type of system is presented in Figure 2.8.

In general case, the formula for PFFO can be written as

$$
P_{k\text{-out-of-}n}(t) = \sum_{j=k}^{n} \binom{n}{j} [p(t)]^j [q(t)]^{n-j} = 1 - \sum_{j=0}^{k-1} \binom{n}{j} [p(t)]^j [q(t)]^j,
$$

$$(2.34)$$

where

$$
\binom{n}{j} = \frac{n!}{j! \cdot (n-j)!} = \frac{n \cdot (n-1) \cdots (n-j+1)}{1 \cdot 2 \cdots j}
$$

is a binomial coefficient.

For highly reliable units (when $q \ll 1/n$), one can use an approximation

$$
P_{k\text{-out-of-}n}(t) \approx 1 - \binom{n}{k-1} [q(t)]^{n-k+1}. \tag{2.35}
$$

If units have exponential distribution of TTF, then using arguments analogous to those used for (2.30), one gets

$$
T_{k\text{-out-of-}n} = \frac{1}{\lambda} \sum_{i=k}^{n} \frac{1}{i}. \tag{2.36}
$$

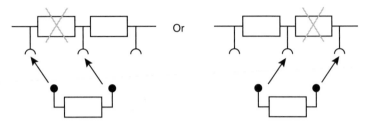

**FIGURE 2.9**    Connection of a redundant unit instead of a failed operating unit.

Note that when $k = n$, the structure transforms into an ordinary series system of $n$ units, and when $k = 1$, it transforms into an ordinary parallel system.

As a rule, structures "2-out-of-3" are found in engineering practice. For such a system, the Boolean expression has the form

$$\phi(\mathbf{x}) = (x_1 \cap x_2 \cap x_3) \cup (\bar{x}_1 \cap x_2 \cap x_3) \cup (x_1 \cap \bar{x}_2 \cap x_3)$$
$$\cup (x_1 \cap x_2 \cap \bar{x}_3). \tag{2.37}$$

The RBD for such a system is presented in Figure 2.9.

From (2.34) for identical units follows

$$P_{2\text{-out-of-3}} = E\{\phi(\mathbf{x})\} = p^3 + 3p^2 q. \tag{2.38}$$

## 2.5   REALISTIC MODELS OF LOADED REDUNDANCY

On the paper, redundant systems are sufficiently reliable. However, a real life is harder than a paper project.

Usually, one should use monitoring of an operating unit for switching to a redundant one after a failure. In addition, there should be a switching device, among others. Thus, the main problem is in constructing a realistic mathematical model. There is no universal solution: a reliability engineer has to take into account all specific features of analyzed equipment and construct an individual mathematical model for it.

We begin with a dubbed system, one of the most cited cases of redundancy in engineering practice. We will consider this case in more detail, taking into account some realistic assumptions.

### 2.5.1 Unreliable Switching Process

What happens if a switching process is unreliable? How much will it affect the dubbed system reliability?

Let us denote probability of successful switching by $\pi$. Then PFFO of an unrecoverable dubbed system can be written as

$$P(t) = (1 - \pi)p(t) + \pi\{1 - [q(t)]^2\}. \qquad (2.39)$$

We would like to demonstrate numerical results using for this purpose a simple model realized in MS Excel. In this example, TTF distribution is taken exponential (Figure 2.10).

### 2.5.2 Non-Instant Switching

Usual assumption that switching from a failed operational unit to a redundant one is instantaneous is erroneous. Actually, many systems

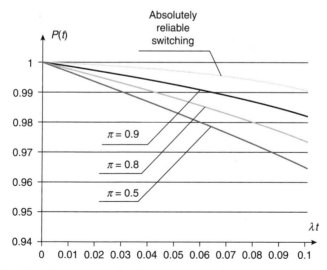

**FIGURE 2.10**  Dependence of a dubbed unrecoverable system on reliability of switching.

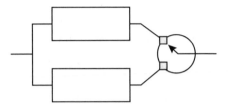

**FIGURE 2.11**  Conditional RBD of an unrecoverable dubbed system with an unreliable switch.

have some "functional inertia": it can stand short downtimes. In other words, there is some "acceptable" switching time, $\varepsilon$, which does not lead to the dubbed system failure.

Switching time itself can be random with some d.f., $F_{\text{switch}}(t)$. In this case, with probability $F_{\text{switch}}(\varepsilon)$ switching is considered as successful and with probability $1 - F_{\text{switch}}(\varepsilon)$ system fails.

It is clear that this case differs from the previous model only by terminology and notation.

### 2.5.3  Unreliable Switch

Now assume that switching process is ideal; however, a switch itself is unreliable and can fail with time. Thus, if the switch has failed before an operating unit failure, then utilization of a redundant unit will be impossible. Let us present a conditional RBD for this case in the form shown in Figure 2.11.

Let us first find the probability, $p^*$, that the switch has failed after the operating unit failed; that is, switching to a redundant unit is successful.

$$p^* = P\{\xi_{\text{switch}} > x|t\} = \int_0^t P_{\text{switch}}(t) \cdot f(t)dt, \qquad (2.40)$$

where $\xi$ is the operating unit TTF, $\xi_{\text{switch}}$ is the switch TTF, $P_{\text{switch}}(t)$ is the switch PFFO, and $f(t)$ is the density function of unit TTF.

Assuming that all d.f.'s are exponential, it is easy to find

$$p^* = \int_0^t \exp(-\lambda_{\text{switch}}t) \cdot \lambda \cdot \exp(-\lambda t)dt = \frac{\lambda}{\lambda_{\text{switch}} + \lambda}. \qquad (2.41)$$

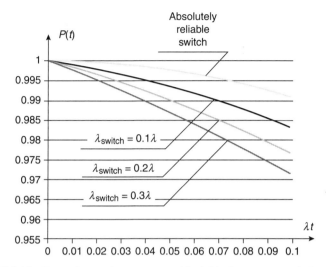

**FIGURE 2.12**    Dependence of an unrecoverable dubbed system on switch reliability.

So, the PFFO of such a system can be written as

$$P(t) = p(t) + q(t) \cdot p^* \cdot p(t). \tag{2.42}$$

Figure 2.12 provides a numerical illustration of a switch unreliability influence.

### 2.5.4    Switch Serving as Interface

Often (especially, in computer systems), the switch plays a role of a special kind of interface between the redundant group and the remaining part of the system. It means that such a switch interface is a necessary unit of the system and, actually, has to be considered as a series unit to the dubbed system. Conditional RBD for this case is presented in Figure 2.13.

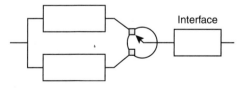

**FIGURE 2.13**    Conditional RBD of an unrecoverable dubbed system with a switch interface.

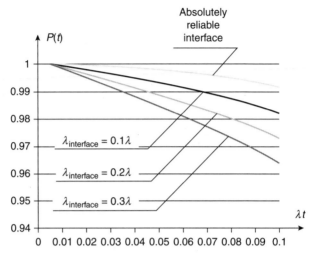

**FIGURE 2.14**   Dependence of an unrecoverable dubbed system on switch interface reliability.

Here we assume that switching is ideal. The switch failure becomes apparent only at the moment of required switch after an operating unit failure. In this case, the PFFO can be calculated by the following formula:

$$P(t) = P_{\text{switch}}(t) \cdot \{1 - [q(t)]^2\}. \tag{2.43}$$

Again for simplicity of numerical calculations assume that all distributions are exponential (Figure 2.14).

From this example, one can see that the switch interface reliability plays a crucial role. Moreover, if switch interface reliability is comparable with unit reliability, then duplication almost has no practical sense.

All these models are given to demonstrate how important can be "secondary" factors, concerning switching from a failed unit to a redundant one.

### 2.5.5   Incomplete Monitoring of the Operating Unit

However, switching is not the only important factor when one deals with the redundancy group analysis. Also, the monitoring procedure is very important: without determination of operating unit failure, it is impossible to make a decision "to switch or not to switch."

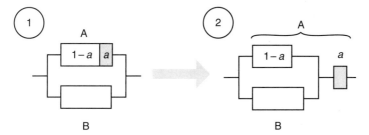

**FIGURE 2.15** Initial RBD of a loaded dubbed system with incomplete monitoring (1) and its equivalent presentation (2).

Assume that some part of the operating unit, say, $a$ ($a < 100\%$), is not controlled at all; that is, any failure of this part becomes a hidden failure, and switching to a redundant unit does not occur. Conditional RBD for this case is presented in Figure 2.15.

The PFFO of such a system can be obtained by using the following formula:

$$P(t) = p_a(t) \cdot [1 - q_{1-a}(t) \cdot q(t)], \qquad (2.44)$$

where $p_a(t)$ is the PFFO of a noncontrolled part of the operating unit and $q_{1-a}(t)$ is failure probability of a controlled part of the operating unit. Figure 2.16 provides results of numerical calculations.

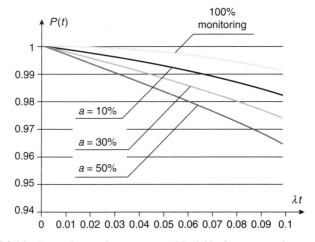

**FIGURE 2.16** Dependence of an unrecoverable dubbed system on the operating unit monitoring completeness.

### 2.5.6   Periodical Monitoring of the Operating Unit

For an unrecoverable redundant system, periodical monitoring has no sense at all: any failure is detected "postmortem," when an operating unit has already failed but switching has not occurred.

## 2.6   REDUCIBLE STRUCTURES

Pure series and pure parallel systems are not seen in engineering practice so often. In general case, systems have more complex structures. However, most of such structures can be reduced to a conditional unit by some simple procedures. Such systems are called reducible.

### 2.6.1   Parallel-Series and Series-Parallel Structures

The simplest reducible systems are parallel-series and series-parallel structures presented in Figure 2.17a and b, respectively.

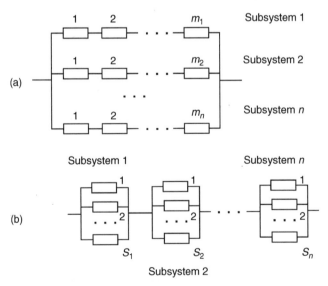

**FIGURE 2.17**   RBDs for (a) parallel-series and (b) series-parallel systems.

We will write only expressions for PFFO without trivial explanations:

$$1. \quad P = 1 - \prod_{j=1}^{n}\left[1 - \prod_{k=1}^{m_j} p_k\right] \qquad (2.45)$$

and

$$2. \quad P = \prod_{j=1}^{n}\left[1 - \prod_{k=1}^{s_j}(1 - p_k)\right]. \qquad (2.46)$$

Of course, such idealized systems are also seldom seen in engineering practice.

### 2.6.2    General Case of Reducible Structures

Avoiding general consideration, let us demonstrate the procedure of reducing on a simple particular RBD (Figure 2.18).

At the beginning, we distinguish series structure (units 2 and 3) and parallel structure (units 4 and 5), and form new "equivalents" 6 and 7.

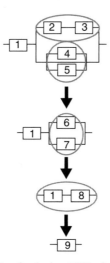

**FIGURE 2.18**    Example of reducing RBD of a system to a single unit.

Then, units 6 and 7 are transformed into unit 8. And finally, we get a single equivalent unit 9.

Construction of the expression for system PFFO starts from the bottom of the scheme of transformation:

$$\begin{aligned}
P = p_9 &= p_1 \cdot p_8 = p_1 \cdot (1 - q_6 q_7) \\
&= p_1 \cdot \{1 - (1 - p_2 p_3) \cdot [1 - (1 - q_4 q_5)]\}.
\end{aligned} \qquad (2.47)$$

Of course, not all structures are reducible; some of them will be considered later.

## 2.7    STANDBY REDUNDANCY

Most technical systems have spare parts that can almost instantaneously replace a failed operating unit. Of course, in practice such a replacement takes time, although most mathematical models assume that the time of replacement is equal to 0.

This type of redundancy is called standby redundancy. In this case, redundant units are not included in an "active" system's structure. Moreover, these redundant units cannot fail until they occupy an active position. Of course, redundant units have to be identical to operating ones by all parameters, including reliability indices.

### 2.7.1    Simple Redundant Group

A system consisting of a single operating unit and $(n-1)$ standby units is called a redundant group. Conditional RBD of a redundant group with standby redundancy can be presented in the form shown in Figure 2.19.

In this case, the random time of the system's successful operation $\xi$ is

$$\xi = \sum_{1 \le k \le n} \xi_k. \qquad (2.48)$$

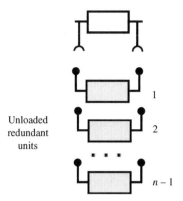

**FIGURE 2.19**   Conditional RBD for a standby redundant group. Standby units are dotted.

Thus, a system's MTTF can be written immediately:

$$T = E\{\xi\} = E\left\{\sum_{1 \leq k \leq n} \xi_k\right\} = \sum_{1 \leq k \leq n} E\{\xi_i\} = \sum_{1 \leq k \leq n} T_k = nT^*, \quad (2.49)$$

where $T^*$ is the single unit's MTTF.

Remember the well-known property of the mean: formula (2.49) is valid even if the standby units are dependent.

The probability of a system's successful operation $P(t)$ can be written as

$$P(t) = \Pr\{\xi \geq t\} = \Pr\left\{\sum_{k=1}^{n} \xi_k \geq t\right\}. \quad (2.50)$$

It is known that the distribution of the sum of random variables is calculated as a convolution by the formula

$$P^{(n)}(t) = 1 - F^{*n}(t) = \int_0^t P^{(n-1)}(t - x)\mathrm{d}F(x), \quad (2.51)$$

where $P^{(k)}(t)$ is the PFFO of the system with $(k-1)$ standby units ($k$ units in the entire redundant group).

Formula (2.51) factually gives only idea of calculation, since in most practical cases only numerical calculations are applicable.

However, in engineering practice, especially for electronic devices, the most frequently used distribution $F(t)$ is exponential. The standby group's random TTF has the Erlang[7] d.f. of the $n$th order (see Appendix A.2.5), and the probability of a failure-free operation is

$$P(t) = \Pr\left\{ \sum_{1 \le k \le n} \xi_k \ge t \right\} = \sum_{0 \le k \le n} \frac{(\lambda t)^k}{k!} e^{-\lambda t}$$

$$= 1 - \sum_{n+1 \le k < \infty} \frac{(\lambda t)^k}{k!} e^{-\lambda t}. \tag{2.52}$$

For $\lambda t \ll 1$, the approximation can be written as

$$P(t) \approx 1 - \frac{(\lambda t)^{n+1}}{(n+1)!}. \tag{2.53}$$

In conclusion, note that standby redundancy is more effective than loaded redundancy. This follows from a simple fact that

$$\xi_{\text{standby}} = \sum_{1 \le k \le n} \xi_k \ge \max_{1 \le k \le n} \xi_k = \xi_{\text{loaded}}. \tag{2.54}$$

The equality is never attained because of the strongly positive values of $\xi$'s.

Of course, the reader should keep in mind that standby redundancy, in practice, demands some time to switch a unit into an active regime. More reasonable is considering unloaded redundancy as spare parts.

---

[7] Agner Krarup Erlang (1878–1929) was a Danish mathematician, statistician, and engineer who invented the fields of traffic engineering and queuing theory. Erlang also created the field of telephone network analysis.

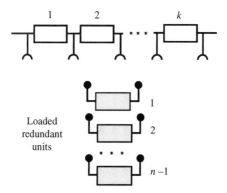

**FIGURE 2.20**    Conditional RBD for a standby "$k$-out-of-$n$" redundant group.

### 2.7.2    Standby Redundancy of Type "$k$-out-of-$n$"

This type of redundancy can be presented by conditional RBD, depicted in Figure 2.20.

It is clear that pure standby redundancy hardly can be implemented in a real technical system. Mostly this type of model is used for evaluation of spare unit's sufficiency. In this case, all units of a series system play the same role within the analyzed equipment.

Consider a series system of $k$ operating units. The system is supported by $(n - k)$ standby units that can replace any failed unit of the group of $k$. In general case, formulas for PFFO and MTTF cannot be written in a simple closed form except for the case of an exponentially distributed random TTF of units. We may write the result basing our explanation on simple arguments.

#### 2.7.2.1    *Mean Time to Failure*    Recall again that we assume that the units are i.i.d.

The system failure rate equals $k\lambda$. There are $(n - k)$ possible replacements, with average interval $1/k\lambda$ between them. So the system MTTF, that is, average time until stock's depletion, is equal to $(n - k + 1)/k\lambda$.

#### 2.7.2.2    *Probability of Failure-Free Operation*    The probability of a system's successful operation when its units have exponential TTF is

described by Poisson[8] distribution:

$$P(t) = \sum_{i=0}^{n-k} \frac{(k\lambda t)^i}{i!}.\qquad(2.55)$$

## 2.8   REALISTIC MODELS OF UNLOADED REDUNDANCY

An ideal model of standby redundancy is constructed under assumption that a standby unit is immediately switched in place of a failed operating unit. This is practically impossible: there should be some kind of functional inertia; that is, the system has to assume a possibility of short downtimes needed for switching. Moreover, here (as well as in the case with loaded redundancy) there are many additional factors influencing a standby redundant group.

Just for simplicity of some expressions, let us assume that units have exponential distribution of TTF. In this case, simple and understandable formulas can be written.

### 2.8.1   Unreliable Switching Process

If switching is ideal, then for an unrecoverable unloaded dubbed system PFFO is written as $P(t) = e^{-\lambda t}(1 + \lambda t)$. Assume that successful switching occurs with probability $\pi$. In this case, system PFFO, $P(t)$, can be calculated as

$$P(t) = (1 - \pi) \cdot e^{-\lambda t} + \pi \cdot e^{-\lambda t}(1 + \lambda t) \approx e^{-\lambda t}(1 + \lambda t).\quad(2.56)$$

How serious is the resulting error due to assumption concerning ideal switching? Let us give numerical illustration, using a simple model based on MS Excel (Figure 2.21).

---

[8] Siméon Denis Poisson (1781–1840) was a French mathematician, geometer, and physicist. He made outstanding contribution in probability theory and theory of stochastic processes.

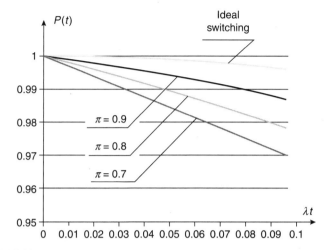

**FIGURE 2.21**    Dubbed system's PFFO depending on probability of switching failure.

## 2.8.2    Non-Instant Switching

Assume that acceptable switching time is equal to $\varepsilon$; that is, if the switching duration $\omega$ is less than $\varepsilon$ ($\omega < \varepsilon$), the system does not fail. If this switching time is a random value with a known distribution function, $F_{\text{switch}}(t)$, the probability of acceptable switching time, $\pi$, can be easily calculated: $\pi = F_{\text{switch}}(\varepsilon)$.

After this, one can use the results of the previous section.

## 2.8.3    Unreliable Switch

Assume that the switching procedure is ideal but a switch itself can fail with time. The RBD of such a system is presented in Figure 2.22.

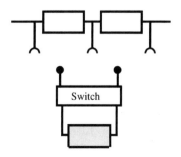

**FIGURE 2.22**    Unloaded redundancy with an unreliable switch.

If the switch has failed before the operating unit, use of a spare unit becomes impossible.

Probability, $p^*$, that the switching device has failed after an operating unit is

$$p^* = P\{\xi_{\text{switch}} > x|t\} = \int_0^t P_{\text{switch}}(t)f(t)dt, \qquad (2.57)$$

where $\xi$ is the unit TTF, $\xi_{\text{switch}}$ is the switch TTF, $P_{\text{switch}}(t)$ is the switch PFFO, and $f(t)$ is the density function of unit TTF.

Assuming that all random variables have exponential distribution, one finds by simple integration:

$$p^* = \int_0^\infty \exp(-\lambda_{\text{switch}}t) \cdot \lambda \cdot \exp(-\lambda t)dt = \frac{\lambda}{\lambda + \lambda_{\text{switch}}}. \qquad (2.58)$$

Again we can get expression for PFFO, substituting $p^*$ instead of $\pi$ in formula (2.56):

$$P(t_0) = p(t_0) + q(t_0) \cdot p^* \cdot p(t_0).$$

Under assumption of exponential distribution of all TTFs, the formula for PFFO can be easily written as

$$P(t) = \frac{\lambda_{\text{switch}}}{\lambda + \lambda_{\text{switch}}} \cdot e^{-\lambda t} + \frac{\lambda}{\lambda + \lambda_{\text{switch}}} \cdot e^{-\lambda t} \cdot (1 + \lambda t)$$
$$= e^{-\lambda t} \cdot \left(1 + \frac{\lambda^2 t}{\lambda + \lambda_{\text{switch}}}\right). \qquad (2.59)$$

This expression has a clear sense: with probability $\lambda_{\text{switch}}/(\lambda + \lambda_{\text{switch}})$ a switch has failed before an operating unit failure, and with probability $\lambda/(\lambda + \lambda_{\text{switch}})$ a switch performs hooking up of a redundant unit. Figure 2.23 numerically illustrates the influence of the switch reliability on the system reliability.

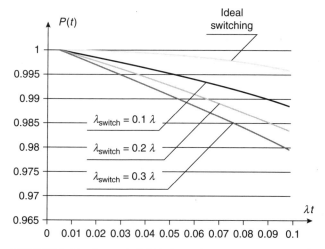

**FIGURE 2.23**   System's PFFO depending on switch reliability.

## 2.8.4   Switch Serving as Interface

In some situations, a switching device is a necessary part of the system operation. For instance, it can be used as an interface between the redundant group and the remaining part of the system. In this case, a switching device has to be considered as a series unit.

Conditional RBD of such a system is presented in Figure 2.24.

The system fails if the redundant group has failed or if the switching device has failed. In this case,

$$
\begin{aligned}
P(t) &= \exp(-\lambda_{\text{switch}} t) \cdot (1 + \lambda t) \cdot e^{-\lambda t} \\
&= \exp[-(\lambda_{\text{switch}} + \lambda)t] \cdot (1 + \lambda t).
\end{aligned}
\tag{2.60}
$$

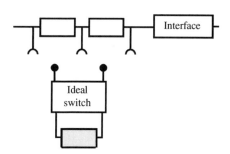

**FIGURE 2.24**   Conditional RBD for unloaded duplication with a switch interface.

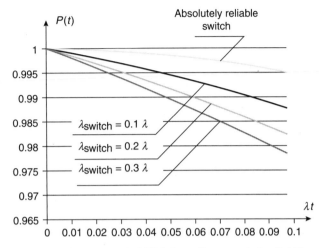

**FIGURE 2.25**    System's PFFO depending on switch reliability.

(We assume that the switching procedure itself is ideal.)

Dependence of system PFFO on parameters of a unit and unreliability of a switch interface is presented in Figure 2.25.

These graphs show how important is the role of switching device reliability. Indeed, a dubbed system cannot be more reliable than the switch interface.

### 2.8.5    Incomplete Monitoring of the Operating Unit

As we saw above, monitoring of the operating system is a very significant factor for unrepairable redundant systems.

Assume that a part, say, $a$ ($a < 100\%$), of an operating unit is not monitoring; that is, its failure leads to the system failure, since there is no indication for switching to a standby unit.

Conditional RBD for this case is presented in Figure 2.26.

PFFO of this system is expressed as

$$P(t) = \Pr\{\xi_a < \xi_{1-a}\} \cdot \Pr\{\xi_a > t\} + \Pr\{\xi_{1-a} < \xi_a\} \cdot [\Pr\{\xi_{1-a} > t\}$$

$$+ \Pr\{\xi_{1-a} < t\} \cdot \Pr\{\xi > t - \xi_{1-a}\}]. \qquad (2.61)$$

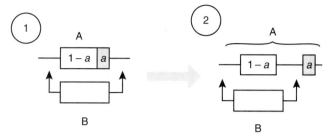

**FIGURE 2.26**  Initial RBD of a standby dubbed system with incomplete monitoring of the operating unit and its equivalent transformation.

In general case, this formula cannot be written in closed form and only numerical integration is possible. However, if all distributions are exponential, one can derive expression for PFFO in closed form:

$$P(t) \approx \frac{\lambda_a}{\lambda} \cdot \exp(-\lambda_a t) + \frac{\lambda_{1-a}}{\lambda} \cdot \exp(-\lambda t) \cdot (1 + \lambda t). \qquad (2.62)$$

Comparison of different variants of monitoring completeness is given in Figure 2.27.

Note that periodical monitoring of an operating unit in the case of standby redundancy also has no sense, since after "hidden" failure the system moves to a failure state.

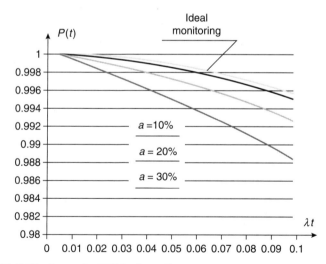

**FIGURE 2.27**  System's PFFO depending on completeness of operating unit monitoring.

# 3

# RECOVERABLE SYSTEMS: MARKOV MODELS

Reliability of recoverable systems with arbitrary distributions of units' time to failure practically cannot be described analytically in a convenient and "workable" form. So, we restricted ourselves to Markov[1] models. By the way, almost all stationary reliability indices for non-Markov models can be derived by substitution of corresponding MTTFs and MDTs in formulas obtained for Markov models.

## 3.1 UNIT

A recoverable unit is defined by its two main parameters—MTTF and MDT. Usually, one assumes that after failure a recovered unit is identical (in statistical sense) to the initial one, so in this case, MTTF and

---

[1] Andrey Andreyevich Markov (1856–1922) was a Russian mathematician. He is best known for his work on theory of stochastic processes. His research later became known as Markov chains.

*Probabilistic Reliability Models*, First Edition. Igor Ushakov.
© 2012 John Wiley & Sons, Inc. Published 2012 by John Wiley & Sons, Inc.

MTBF coincide. We are beginning with the simplest case when both distribution TTF and recovery time are exponential.

### 3.1.1 Markov Model

**3.1.1.1 General Description**   In most academic approaches, random TTF and random recovery time are assumed exponentially distributed that gives a possibility to use the Markov model for reliability study. Let the parameter of the TTF distribution be $\lambda$ and the parameter of the recovery time be $\mu$. In other words, MTBF (MTTF) and mean recovery time are known from the beginning: $T = 1/\lambda$ and $\tau = 1/\mu$.

At any moment of time, the unit can be in one of the two states: either operational or failed. A convenient form of presentation of such a process is the transition graph presented in Figure 3.1. Let us denote the operational state with symbol "0" and the failed state with symbol "1".

The unit transition process can be described as an alternative renewal process. It is represented by a sequence of mutually indepen- dent r.v.'s $\xi_k$ (unit's operational time) and $\eta_k$ (unit's recovery time). An example of a time diagram is presented in Figure 3.2.

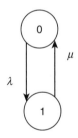

**FIGURE 3.1**   Transition graph for a recoverable unit.

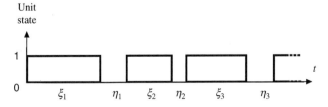

**FIGURE 3.2**   Example of a time diagram for a unit.

### 3.1.1.2 Equations for Finding Nonstationary Availability Coefficient
Let us find the probability, $p_0(t)$, that at moment $t + \Delta t$ a unit is in state "0". There are two possibilities:

- at moment $t$ unit was in state "0" and did not leave it during infinitesimally small time interval $\Delta t$, which happens with probability $1 - \lambda \Delta t$, or
- at moment $t$ it was in state "1" and moved to the state "0" during the time interval $\Delta t$, which happens with probability $\mu \Delta t$.

This immediately gives

$$p_0(t + \Delta t) = (1 - \lambda \Delta t)p_0(t) + \mu \Delta t\, p_1(t). \tag{3.1}$$

From (3.1), we easily obtain

$$\frac{p_0(t + \Delta t) - p_0(t)}{\Delta t} = -\lambda p_0(t) + \mu p_1(t), \tag{3.2}$$

and in the limit as $\Delta t \to 0$, we obtain the following differential equation:

$$\frac{d}{dt} p_0(t) = -\lambda p_0(t) + \mu p_1(t). \tag{3.3}$$

This represents the simplest example of the Chapman[2]–Kolmogorov[3] equation. To solve it with respect to any $p_k(t)$, we need to have one more equation. The second equation, which is called the normalization equation, is

$$p_0(t) + p_1(t) = 1, \tag{3.4}$$

---

[2] Sydney Chapman (1888–1970) was a British mathematician and geophysicist.
[3] Andrey Nikolaevich Kolmogorov (1903–1987) was a great Russian mathematician of the twentieth century who advanced various scientific fields, among them probability theory, topology, intuitionistic logic, turbulence, classical mechanics, and computational complexity.

which means that at any moment the unit must be in one of the two possible states.

We also need to determine the initial condition for the solution of the system of differential equations. Usually, one assumes that at moment $t=0$ the unit is in operational state, that is, $p_0(0)=1$.

It is clear that $p_0(t)$ is a nonstationary availability coefficient that shows the probability that a unit has been found in state "0" at a given moment $t$ under condition that at moment $t=0$ the unit was in state "0".

This reliability index can be found with the help of different methods. We will use the Laplace[4] transform (LT) to make the presentations of other solutions in the book uniform. Brief information about Laplace transforms can be found in Appendix B.

***3.1.1.3 Nonstationary Availability Coefficient*** The system of above differential equations (3.3)–(3.4) with the initial condition $p_0(t)=1$ has the LST form:

$$-1 + s\,\varphi_0(s) = -\lambda\varphi_0(s) + \mu\varphi_1(s),$$
$$\varphi_0(s) + \varphi_1(s) = \frac{1}{s}, \tag{3.5}$$

or, in the canonical form,

$$(\lambda + s)\,\varphi_0(s) - \mu\,\varphi_1(s) = 1,$$
$$s\,\varphi_0(s) + s\,\varphi_1(s) = 1. \tag{3.6}$$

To solve (3.6), we can use Cramer's[5] rule:

$$\varphi_0(s) = \frac{\begin{vmatrix} 1 & -\mu \\ 1 & s \end{vmatrix}}{\begin{vmatrix} \lambda+s & -\mu \\ s & s \end{vmatrix}} = \frac{s+\mu}{s^2+(\lambda+\mu)s}. \tag{3.7}$$

[4] Pierre-Simon, Marquis de Laplace (1749–1827) was a French mathematician and astronomer whose work was pivotal to the development of mathematical astronomy and statistics.
[5] Gabriel Cramer (1704–1752) was a Swiss mathematician. In linear algebra, Cramer's rule is a theorem that gives an expression for the solution of a system of linear equations in terms of the determinants.

To invert this LST, we have to present it in a form of a sum of terms of types $a/s$ or $b/(s + a)$, inverse functions for which are a constant and an exponential function, respectively.

The denominator of fraction in (3.7) can be written as $s^2 + (\lambda + \mu)s = (s - s_1)(s - s_2)$, where $s_1$ and $s_2$ are polynomial roots that are, as can be easily found, $s_1 = 0$ and $s_2 = -(\lambda + \mu)$. Now we can write

$$\varphi_0(s) = \frac{A}{s - s_1} + \frac{B}{s - s_2} = \frac{A}{s} + \frac{B}{s + \lambda + \mu}, \tag{3.8}$$

where $A$ and $B$ are unknown constants to be determined. To find them, we should note that two polynomials with similar denominators are equal if and only if the coefficients of their numerators are equal. Thus, we set the two representations equal:

$$\frac{A}{s} + \frac{B}{\lambda + \mu + s} = \frac{s + \mu}{s(\lambda + \mu + s)}. \tag{3.9}$$

And so we obtain a new system for $A$ and $B$ by equalizing the coefficients of the polynomials:

$$\begin{aligned} A + B &= 1, \\ A(\lambda + \mu) &= \mu. \end{aligned} \tag{3.10}$$

It is easy to find

$$\begin{aligned} A &= \frac{\mu}{\lambda + \mu}, \\ B &= 1 - \frac{\mu}{\lambda + \mu} = \frac{\lambda}{\lambda + \mu}. \end{aligned} \tag{3.11}$$

Thus, the LST of interest can be written as

$$\varphi_0(s) = \frac{\mu}{\lambda + \mu} \cdot \frac{1}{s} + \frac{\lambda}{\lambda + \mu} \cdot \frac{1}{\lambda + \mu + s}. \tag{3.12}$$

Finally, the nonstationary availability coefficient, that is, the inverse LST of (3.12), is

$$K(t) = p_0(t) = \frac{\mu}{\mu + \lambda} + \frac{\lambda}{\lambda + \mu} e^{-(\lambda + \mu)t}. \tag{3.13}$$

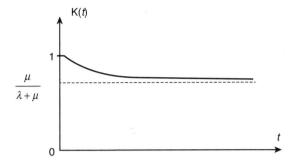

**FIGURE 3.3**    Graph of $K(t)$ with the initial condition $p_0(0) = 1$.

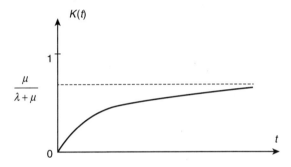

**FIGURE 3.4**    Graph of $K(t)$ with the initial condition $p_1(0) = 1$.

The function $K(t)$ showing the time dependence of the system availability is presented in Figure 3.3.

By the way, if the initial condition is $p_1(0) = 1$, then the graph of $K(t)$ will be as shown in Figure 3.4.

The graph shows that after a while $K(t)$ approaches a stationary value independently of the initial state. Since index $K(t)$ is practically almost never used in practice, we restrict ourselves by considering it for a recoverable unit.

**3.1.1.4    Stationary Availability Coefficient**    As mentioned above, if $t \to \infty$, $K(t)$ approaches its limit value that is called the stationary availability coefficient, $K$:

$$K = \lim_{t \to \infty} K(t) = \frac{\mu}{\lambda + \mu} + \frac{\lambda}{\lambda + \mu} \lim_{t \to \infty} \exp[-(\lambda + \mu)t] = \frac{\mu}{\lambda + \mu}$$
$$= \frac{T}{T + \tau}. \tag{3.14}$$

Actually, the availability coefficient can be defined as an average portion of time, when a unit is in operating state. In turn, this is the average portion of time when a unit is in operating state during a single "operating–recovering" cycle. So, expression (3.14) for a recoverable unit can be written directly from the definition of availability coefficient.

It is time to repeat that we use all this rather sophisticated mathematics solely to demonstrate general methodology on simplest examples.

In practice, one usually considers highly reliable objects, for which condition $E\{\eta\}/E\{\xi\} \ll 1$ is satisfied. In this case, it is possible to use a very good approximation:

$$K = \frac{T}{T+\tau} = \frac{1}{1+\tau/T} \approx 1 - \frac{\tau}{T} = 1 - \frac{\lambda}{\mu}. \qquad (3.15)$$

Error of this approximation does not exceed $(\lambda/\mu)^2$.

### 3.1.1.5 Probability of Failure-Free Operation

Since MTTF is given and the Markov model is considered, one can immediately write

$$P(t) = p_0(t) = e^{-\lambda t}. \qquad (3.16)$$

Note that for a highly reliable unit there is a simple and accurate approximation:

$$P_0(t) \approx 1 - \lambda t. \qquad (3.17)$$

This approximation has an error of order $(\lambda t)^2$.

### 3.1.1.6 Coefficient of Interval Availability

The easiest way to get this reliability index (that we denote by $R(t_0)$) is to use memoryless or Markov property. For this case, we can just multiply availability coefficient by PFFO, that is,

$$R(t_0) = P(t, t_0) \cdot K = P(0, t_0) \cdot K = \frac{T}{T+\tau} \exp(-\lambda t_0). \qquad (3.18)$$

For highly reliable systems, one can write an approximation

$$R(t_0) \approx 1 - \lambda(t_0 + \tau). \tag{3.19}$$

*Remark.* We analyzed this simple case with such a scrupulosity only to demonstrate different possible ways of obtaining the needed result. We do this to avoid explanations below with unnecessary additional details for more complex models. The same purpose drives us to use a homogeneous mathematical technique for all routine approaches.

## 3.2  SERIES SYSTEM

Recoverable series systems differ by their recovery processes. First of all, some systems have to be turned off during recovery after failure. In this case, there is a single failed unit under restoration. Another case: system continues to stay in an operational state, so during recovering a currently failed unit new failures may appear. In principle, in this case one can observe even a situation when all system units have failed. It can happen if, for instance, a recovery process is very slow.

In addition, the number of repair facilities can be restricted, so failed units can form a queue for recovering.

### 3.2.1  Turning Off System During Recovery

Assume that distributions of the TTF, $F_i(t)$, and of the recovery time, $G_i(t)$, are exponential for all units. Denote parameters of these d.f.'s by $\lambda_i$ and $\mu_i$, respectively.

Let after failure of any unit, the system is turned off during recovery of the failed unit, so other units cannot fail until recovery completion. The transition graph for such a system is presented in Figure 3.5.

We will not write the equations to obtain results for this case. As much as possible, we will try to use simple verbal explanations.

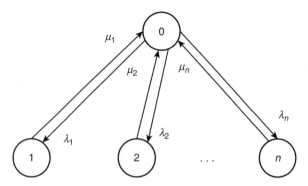

**FIGURE 3.5**    Transition graph for a series system that is turned off during recovery.

### 3.2.1.1  Probability of Failure-Free Operation    Any exit from state "0" (see Figure 3.5) leads to failure. Hence,

$$P(t) = \exp\left(-\sum_{1 \leq i \leq n} \lambda_i t\right). \tag{3.20}$$

Thus, by its PFFO the system is equivalent to a single unit with a failure rate equal to $\Lambda$:

$$\Lambda = \sum_{1 \leq i \leq n} \lambda_i. \tag{3.21}$$

### 3.2.1.2  Mean Time to Failure    From (3.21), one easily finds that

$$T = \frac{1}{\displaystyle\sum_{1 \leq i \leq n} \lambda_i}. \tag{3.22}$$

If all system units have exponential distributions of TTF, then the systems MTTF and MTBF coincide.

### 3.2.1.3  Mean Recovery Time    Let us consider a general case, where all units differ by their repair time $\tau_i = 1/\mu_i$. It is clear that a current system's failure due to unit $i$ is

$$p_i = \frac{\lambda_i}{\Lambda}, \tag{3.23}$$

where $\Lambda$ is defined in (3.21).

Thus, mean recovery time can be easily found as a weighed value:

$$\tau = \sum_{1 \le i \le n} \frac{p_i}{\mu_i} = \frac{1}{\Lambda} \sum_{1 \le i \le n} \frac{\lambda_i}{\mu_i}. \tag{3.24}$$

### 3.2.1.4 Stationary Availability Coefficient
Using (3.21) and (3.24), one easily writes

$$K = \frac{1}{1 + \sum_{1 \le i \le n} \lambda_i \tau_i} \approx 1 - \sum_{1 \le i \le n} \lambda_i \tau_i. \tag{3.25}$$

It is important to note that if distributions $F_i(t)$ and $G_i(t)$ are not exponential, the expression

$$K = \frac{1}{1 + \sum_{1 \le i \le n} \tau_i / T_i} \approx 1 - \sum_{1 \le i \le n} \frac{\tau_i}{T_i} \tag{3.26}$$

remains valid. (Conditions of high reliability for the approximation correctness conserve.)

### 3.2.1.5 Stationary Interval Availability Coefficient
Since distribution of the system's TTF is exponential, we can use the expression $R(t_0) = KP(t_0)$, where $P(t_0)$ and $K$ are defined in (3.20) and (3.25), respectively.

## 3.2.2 System in Operating State During Recovery: Unrestricted Repair

Consider a recoverable series system of $n$ independent units with $n$ independent repair facilities for a case when the system continues to stay in up state, so any of its unit may fail during the recovery process of the previously failed unit.

In this case, the system's reliability indices can be obtained in a very simple way.

### 3.2.2.1 Probability of Failure-Free Operation and Mean Time to Failure
The system PFFO and MTTF coincide with those considered above in (3.20) and (3.22). It is also clear that MTBF is equal to MTTF, since all units have exponential distribution of TTF.

### 3.2.2.2 Mean Recovery Time

Let us consider a general case where each unit has its own mean repair time $\tau_i = 1/\mu_i$. It is clear that a current system's failure due to unit $i$ occurs with probability

$$p_i = \frac{\lambda_i}{\Lambda}, \tag{3.27}$$

where $\Lambda$ is the total system failure rate defined in (3.21).

Thus, mean recovery time can be easily found as a weighed value:

$$\tau = \sum_{1 \le i \le n} \frac{p_i}{\mu_i} = \frac{1}{\Lambda} \sum_{1 \le i \le n} \frac{\lambda_i}{\mu_i}. \tag{3.28}$$

Note that recovery time has a hyperexponential distribution (see Appendix A.2.6):

$$\Pr\{\eta_{\text{Cucm.}} \ge t\} = \frac{\lambda_i \exp(-\lambda_i t)}{\sum_{i=1}^{n} \lambda_i}. \tag{3.29}$$

### 3.2.2.3 Stationary Availability Coefficient

For independent units, one immediately writes

$$K = \prod_{1 \le i \le n} \frac{1}{1 + \lambda_i \tau_i} \approx 1 - \sum_{1 \le i \le n} \lambda_i \tau_i. \tag{3.30}$$

(Conditions of approximation correctness are the same as above in analogous cases.)

### 3.2.2.4 Stationary Operational Availability Coefficient

Since distribution of the system TTF is exponential, we can use the expression $R(t_0) = KP(t_0)$, where $P(t_0)$ and $K$ are defined in (3.20) and (3.25), respectively.

*Remark.* If the number of independent repair facilities $k$ is less than $n$, solution for different units becomes very clumsy and actually has pure "academic" interest.

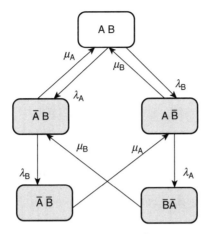

**FIGURE 3.6**  Example of a transition graph for a series system of two units with a single repair facility.

Just for demonstration of the above-mentioned fact, let us consider a relatively simple system of two different independent units and one repair facility. Assume that repair of failed units conforms to the rule "first-in, first-out." (In Figure 3.6, system's failure states are shadowed.)

### 3.2.3  System in Operating State During Recovery: Restricted Repair

Sometimes when repair facilities are restricted in their ability to simultaneously recover several failures, a queue of failed units can be formed. Naturally, it leads to the recovery time increase. Analysis of such systems in general case cannot lead to obtaining convenient formulas. However, if all units of a series system are assumed identical (rather rare case in real engineering practice!), the problem of finding reliability indices becomes solvable. In this case, one can use birth-and-death model described in Appendix C.2. In this particular case, when there are only $k$ repair facilities ($k < n$) for a series system of $n$ units, the transition graph has the form depicted in Figure 3.7.

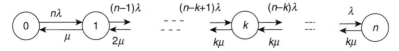

**FIGURE 3.7**  Transition graph for a system of $n$ identical units with $k$ repair facilities.

In the considered case, the system of equations is

$$n\lambda p_0 = \mu p_1,$$
$$(n-1)\lambda p_1 = 2\mu p_2,$$
$$\vdots$$
$$(n-k+1)\lambda p_{k-1} = k\,\mu p_k, \qquad (3.31)$$
$$(n-k)\,\lambda p_k = k\,\mu p_{k+1},$$
$$\vdots$$
$$\lambda p_{n-1} = k\,\mu p_n.$$

Standard solution of the system (3.31), given in Appendix C.1.4, has the form

$$p_1 = n\rho p_0 = \binom{n}{1}\rho p_0,$$
$$p_2 = \frac{n-1}{2}\rho p_1 = \frac{n(n-1)}{2}\rho^2 p_0 = \binom{n}{2}\rho^2 p_0,$$
$$\vdots$$
$$p_k = \frac{n-k+1}{k}\rho p_{k-1} = \frac{n(n-1)\cdots(n-k+1)}{1\cdot 2\cdots k}\rho^k p_0 = \binom{n}{k}\rho^k p_0,$$
$$p_{k+1} = \frac{n-k}{k}\rho p_k = \frac{n(n-1)\cdots(n-k)}{1\cdot 2\cdots k\cdot k}\rho^k p_0 = \binom{n}{k-1}\frac{n-k}{k}\rho^{k+1} p_0,$$
$$\vdots$$
$$p_n = \binom{n}{k}\frac{(n-k)!}{k^{n-k}}\rho^n p_0, \qquad (3.32)$$

where $\rho = \lambda/\mu$.

Since sum of all these probabilities is equal to 1 (condition of the total probability), from (3.32), one easily finds the stationary availability coefficient:

$$K = p_0 = \left[ 1 + \sum_{i=1}^{k}\binom{n}{i}\rho^i + \binom{n}{k}\rho^k \sum_{j=0}^{n-k}\frac{(n-k)!}{j!k^i}\rho^j \right]^{-1}. \qquad (3.33)$$

If the considered system is highly reliable, that is, $\rho \ll 1/n$, one can write an approximation

$$K \approx 1 - n\rho. \qquad (3.34)$$

Actually, it says that for highly reliable systems, it is enough to have a single repair facility because the probability of occurrence of another failure during recovery time is infinitesimally small.

## 3.3  DUBBED SYSTEM

We begin with the particular case of parallel systems because it allows demonstration of mathematical technique on a clear and understandable level.

### 3.3.1  General Description

A dubbed recoverable system with embedded loaded redundant unit is probably the most common case of redundancy in engineering practice. This simple structure allows us to perform a general analysis for all possible configurations: loaded and unloaded redundancy for restricted and unrestricted number of repair facilities. Transition graphs for all these cases are presented in Figure 3.8.

Usually, one makes the following assumptions:

1. In case of loaded redundancy, failures of both units occur independently.
2. After failure of an operating unit, switching to a redundant unit is instantaneous and absolutely reliable.
3. Recovering of a failed unit begins immediately if repairing resources are available.
4. After recovering, a unit becomes as well as the initial one.

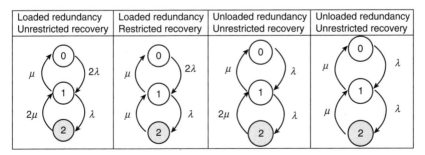

| Loaded redundancy Unrestricted recovery | Loaded redundancy Restricted recovery | Unloaded redundancy Unrestricted recovery | Unloaded redundancy Unrestricted recovery |
|---|---|---|---|

**FIGURE 3.8**    Transition graphs for four cases of recoverable dubbed systems.

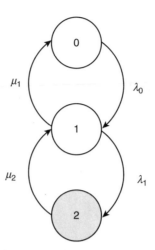

**FIGURE 3.9**  Transition graph for the general Markov model of recoverable dubbed systems.

We will find solution of the problem in general case, using the transition graph depicted in Figure 3.9.

### 3.3.2  Nonstationary Availability Coefficient

Let us write a system of differential equations in the same way as we did for a single recoverable unit:

$$\frac{d}{dt}p_0(t) = -\lambda_0 p_0(t) + \mu_1 p_1(t),$$

$$\frac{d}{dt}p_1(t) = \lambda_0 p_0(t) - (\lambda_1 + \mu_1)p_1(t) + \mu_2 p_2(t), \qquad (3.35)$$

$$1 = p_0(t) + p_1(t) + p_2(t),$$

$$p_0(0) = 1.$$

Laplace transform gives us the following system of algebraic equations:

$$-1 + s\varphi_0(s) = -\lambda_0 \varphi_0(s) + \mu_1 \varphi_1(s),$$

$$s\varphi_1(s) = \lambda_0 \varphi_0(s) - (\lambda_1 + \mu_1)\varphi_1(s) + \mu_2 \varphi_2(s), \qquad (3.36)$$

$$\frac{1}{s} = \varphi_0(s) + \varphi_1(s) + \varphi_2(s).$$

Since for a dubbed system both states "0" and "1" are operational, expression for the availability coefficient can be written as

$$K(t) = p_0(t) + p_1(t) = 1 - p_2(t), \qquad (3.37)$$

that is, LST of function $K(t)$ is

$$\varphi_K(s) = 1/s - \varphi_2(s). \qquad (3.38)$$

To find $\varphi_K(s) = 1/s - \varphi_2(s)$ from system of equation (3.36), we can apply Cramer's rule:

$$\varphi_2(s) = \frac{\begin{vmatrix} \lambda_0 + s & -\mu_1 & 1 \\ -\lambda_0 & \lambda_1 + \mu_1 + s & 0 \\ s & s & 1 \end{vmatrix}}{\begin{vmatrix} \lambda_0 + s & -\mu_1 & 0 \\ -\lambda_0 & \lambda_1 + \mu_1 + s & -\mu_2 \\ s & s & s \end{vmatrix}}$$

$$= \frac{\lambda_0 \lambda_1}{s[s^2 + s(\lambda_0 + \lambda_1 + \mu_1 + \mu_2) + \lambda_0 \lambda_1 + \lambda_0 \mu_2 + \mu_1 \mu_2]}. \qquad (3.39)$$

Finally, expression for the needed LST is

$$\varphi_K(s) = \frac{1}{s} - \varphi_2(s)$$

$$= \frac{s^2 + s(\lambda_0 + \lambda_1 + \mu_1 + \mu_1) + \lambda_0 \mu_2 + \mu_1 \mu_2}{s[s^2 + s(\lambda_0 + \lambda_1 + \mu_1 + \mu_2) + \lambda_0 \lambda_1 + \lambda_0 \mu_2 + \mu_1 \mu_2]}. \qquad (3.40)$$

For preparing to inverse LST, let us present (3.40) in the form of simple fractions of the following form:

$$\varphi_K(s) = \frac{A}{s - s_1} + \frac{B}{s - s_2} + \frac{C}{s - s_3}, \qquad (3.41)$$

where $s_1$, $s_2$, and $s_3$ are roots of denominator in (3.40). In this case, $s_2$ and $s_3$ are conjugate roots, and $s_1 = 0$:

$$s_{2,3} = -\frac{\alpha}{2} \pm \sqrt{\frac{\alpha^2}{4} - \beta}, \qquad (3.42)$$

where, in turn, $\alpha = \lambda_0 + \lambda_1 + \mu_1 + \mu_2$ and $\beta = \lambda_0\lambda_1 + \lambda_0\mu_2 + \mu_1\mu_2$. (Naturally, if $s_2 = s_3$, one uses l'Hôpital's[6] rule.)

The next step is finding coefficients $A$, $B$, and $C$ in (3.41). First, make in (3.41) reduction to a common denominator:

$$\varphi_K(s) = \frac{A(s - s_2) \cdot (s - s_3) + B(s - s_1) \cdot (s - s_3) + C(s - s_1) \cdot (s - s_2)}{(s - s_1) \cdot (s - s_2) \cdot (s - s_3)}.$$

$$(3.43)$$

Two fractions (3.40) and (3.43) with equal denominators are equal if and only if polynomials in numerators are also equal. This condition permits us to write immediately

$$\begin{aligned} s^2 + s(\lambda_0 + \lambda_1 + \mu_1 + \mu_1) + \lambda_0\mu_2 + \mu_1\mu_2 \\ = A(s - s_2) \cdot (s - s_3) + B(s - s_1) \cdot (s - s_3) + C(s - s_1) \cdot (s - s_2). \end{aligned}$$

$$(3.44)$$

The right-hand side of the equality (3.44) can be rewritten as

$$\begin{aligned} s^2(A + B + C) - s[A(s_2 + s_3) + B(s_1 + s_3) + C(s_1 + s_2)] + As_2s_3 \\ + Bs_1s_3 + Cs_1s_2. \end{aligned}$$

$$(3.45)$$

In result, (3.44) and (3.45) allow us to compile a new system of algebraic equations for finding coefficients $A$, $B$, and $C$:

$$\begin{aligned} A + B + C &= 1, \\ A(\lambda + \mu) - 2B(\lambda + \mu) - C(\lambda + \mu) &= 3\lambda + 3\mu, \\ A(\lambda + \mu)^2 &= 2\lambda^2 + 4\lambda\mu + 2\mu^2. \end{aligned}$$

$$(3.46)$$

We omit simple, however rather boring, mathematical exercises and present the final result for all four different cases of recoverable dubbed systems in Table 3.1.

---

[6] Guillaume François Antoine, Marquis de l'Hôpital (1661–1704) was a French mathematician. l'Hôpital's rule for calculating limits involving indeterminate forms 0/0 and ∞/∞ did not originate with l'Hôpital; it appeared in print for the first time in his famous book, which was the first systematic exposition of differential calculus.

**TABLE 3.1 Formulas of $K(t)$ for Four Cases of Recoverable Dubbed Systems**

Loaded redundancy: unrestricted recovery

$$1 - \frac{2\lambda^2}{\varepsilon_1\varepsilon_2}\left[1 - \frac{1}{\varepsilon_1-\varepsilon_2}\left(\varepsilon_1 e^{-\varepsilon_1 t} - \varepsilon_2 e^{-\varepsilon_2 t}\right)\right] \approx 1 - \gamma^2\left[1 - \left(2 - \exp\left(-\frac{\lambda t}{\gamma}\right)\right)\exp\left(-\frac{\lambda t}{\gamma}\right)\right],$$

where $\varepsilon_{1,2} = \dfrac{\lambda}{2\gamma}(1+\gamma)(3\pm1)$ and $\gamma = \dfrac{\lambda}{\mu}$

Loaded redundancy: restricted recovery

$$1 - \frac{2\lambda^2}{\varepsilon_1\varepsilon_2}\left[1 - \frac{1}{\varepsilon_1-\varepsilon_2}\left(\varepsilon_1 e^{-\varepsilon_1 t} - \varepsilon_2 e^{-\varepsilon_2 t}\right)\right] \approx 1 - 2\gamma^2\left[1 - \left(1+\frac{\lambda t}{\gamma}\right)\exp\left(-\frac{\lambda t}{\gamma}\right)\right],$$

where $\varepsilon_{1,2} = \dfrac{\lambda}{2\gamma}\left(2+3\gamma\pm\sqrt{4\gamma+\gamma^2}\right)$ and $\gamma = \dfrac{\lambda}{\mu}$

Unloaded redundancy: unrestricted recovery

$$1 - \frac{2\lambda^2}{\varepsilon_1\varepsilon_2}\left[1 - \frac{1}{\varepsilon_1-\varepsilon_2}\left(\varepsilon_1 e^{-\varepsilon_1 t} - \varepsilon_2 e^{-\varepsilon_2 t}\right)\right] \approx 1 - \frac{\gamma^2}{2}\left[1 - \left(2 - \exp\left(-\frac{\lambda t}{\gamma}\right)\right)\exp\left(-\frac{\lambda t}{\gamma}\right)\right],$$

where $\varepsilon_{1,2} = \dfrac{\lambda}{2\gamma}\left(3+2\gamma\pm\sqrt{1+4\gamma}\right)$ and $\gamma = \dfrac{\lambda}{\mu}$

Unloaded redundancy: restricted recovery

$$1 - \frac{2\lambda^2}{\varepsilon_1\varepsilon_2}\left[1 - \frac{1}{\varepsilon_1-\varepsilon_2}\left(\varepsilon_1 e^{-\varepsilon_1 t} - \varepsilon_2 e^{-\varepsilon_2 t}\right)\right] \approx 1 - \gamma^2\left[1 - \left(1+\frac{\lambda t}{\gamma}\right)\exp\left(-\frac{\lambda t}{\gamma}\right)\right],$$

where $\varepsilon_{1,2} = \dfrac{\lambda}{\gamma}(1+\gamma\pm\sqrt{\gamma})$ and $\gamma = \dfrac{\lambda}{\mu}$

Let us note that in this case when both units are mutually indepen-dent (case of loaded redundancy with unrestricted repair facilities), the solution for $K(t)$ can be obtained directly from the definition of a dubbed system of the mentioned type:

$$K(t) = 1 - [1 - K_{\text{unit}}(t)]^2 = 1 - \left[\frac{\lambda}{\lambda + \mu} \cdot e^{-(\lambda + \mu)t}\right]^2$$

$$= 1 - \frac{\lambda^2}{(\lambda + \mu)^2} \cdot e^{=2(\lambda + \mu)t}. \qquad (3.47)$$

*Remark.* Note again that a nonstationary availability coefficient in practice is used extremely rarely, so these deductions were done exclu-sively from methodological viewpoint, rather than for practical purposes.

We omit deriving availability coefficient for other particular cases, since it will be a boring use of the same standard methods. Neverthe-less, it is interesting to compare all four cases.

### 3.3.3  Stationary Availability Coefficient

For finding this reliability index, one can use the equation system (3.35), replacing all derivatives by 0, all $p_k(t)$ by constant $p_k$, and omit-ting the initial condition. Then one gets the following system of alge-braic equations:

$$\begin{cases} 0 = -\lambda_0 p_0 + \mu_1 p_1, \\ 0 = \lambda_0 p_0 - (\lambda_1 + \mu_1)p_1 + \mu_2 p_2, \\ 1 = p_0 + p_1 + p_2. \end{cases} \qquad (3.48)$$

From the first equation, one gets

$$p_0 = \frac{\mu_1}{\lambda_0} p_1. \qquad (3.49)$$

After substitution of (3.49) into the second equation in (3.48), one gets

$$0 = \mu_1 p_1 - (\lambda_1 + \mu_1)p_1 + \mu_2 p_2 = -\lambda_1 p_1 + \mu_2 p_2 \Rightarrow p_1 = \frac{\mu_2}{\lambda_1} p_2.$$

$$(3.50)$$

Finally, after substitution of (3.49) and (3.50) into the third equation in (3.48):

$$p_2 + \frac{\mu_2}{\lambda_1}p_2 + \frac{\mu_1}{\lambda_0} \cdot \frac{\mu_2}{\lambda_1}p_2 = 1 \Rightarrow p_2$$

$$= \frac{1}{1 + (\mu_2/\lambda_1) + (\mu_1/\lambda_0) \cdot (\mu_2/\lambda_1)}$$

$$= \frac{\lambda_0\lambda_1}{\lambda_0\lambda_1 + \lambda_0\mu_2 + \mu_1\mu_2}, \tag{3.51}$$

or

$$K = \left(1 + \frac{\lambda_0\lambda_1}{\lambda_0\mu_2 + \mu_1\mu_2}\right)^{-1}. \tag{3.52}$$

From (3.52), it is easy to compile Table 3.2.

Naturally, availability coefficient for the first case can be obtained directly:

$$K = 1 - (1 - K)^2 = 1 - \left(\frac{\lambda}{\lambda + \mu}\right)^2 = 1 - \frac{1}{[1 + (1/\gamma)]^2}$$

$$= 1 - \frac{\gamma^2}{(1 + \gamma)^2} \approx 1 - \gamma^2. \tag{3.53}$$

### 3.3.4 Probability of Failure-Free Operation

Transition graphs for this case are presented in Figure 3.10. State "2" corresponding to system's failure is absorbing one, since there is no difference between restricted and unrestricted repair.

**TABLE 3.2 Formulas of $K$ for Four Cases of Recoverable Dubbed Systems**

| | | | |
|---|---|---|---|
| Loaded redundancy: unrestricted recovery | $\dfrac{1}{1 + \gamma^*} \approx 1 - \gamma^2,$ | where $\gamma^* = \dfrac{\gamma^2}{1 + 2\gamma}$ | and $\gamma = \dfrac{\lambda}{\mu}$ |
| Loaded redundancy: restricted recovery | $\dfrac{1}{1 + \gamma^*} \approx 1 - 2\gamma^2,$ | where $\gamma^* = \dfrac{2\gamma^2}{1 + 2\gamma}$ | and $\gamma = \dfrac{\lambda}{\mu}$ |
| Unloaded redundancy: unrestricted recovery | $\dfrac{1}{1 + \gamma^*} \approx 1 - \dfrac{\gamma^2}{2},$ | where $\gamma^* = \dfrac{\gamma^2}{2(1 + \gamma)}$ | and $\gamma = \dfrac{\lambda}{\mu}$ |
| Unloaded redundancy: restricted recovery | $\dfrac{1}{1 + \gamma^*} \approx 1 - \gamma^2,$ | where $\gamma^* = \dfrac{\gamma^2}{1 + \gamma}$ | and $\gamma = \dfrac{\lambda}{\mu}$ |

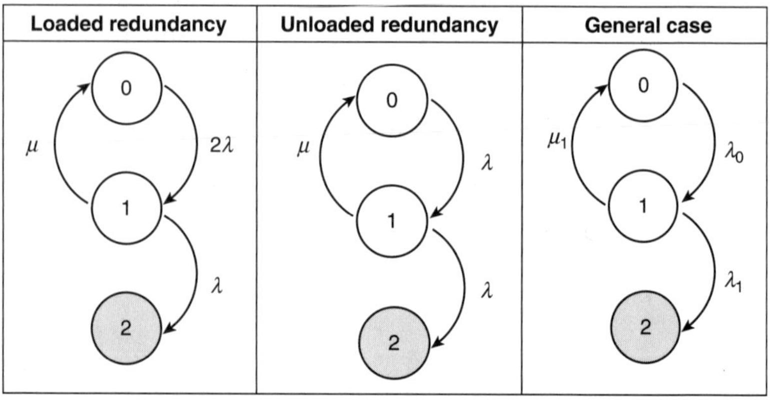
Loaded redundancy
Unloaded redundancy
General case
0
1
2
μ
2λ
λ
μ
λ
λ
μ1
λ0
λ1

**FIGURE 3.10** Transition graphs for calculation of $P(t)$ of recoverable dubbed systems.

The system of linear differential equations in this case is

$$\begin{aligned}\frac{\mathrm{d}}{\mathrm{d}t}p_0(t) &= -\lambda_0 p_0(t) + \mu_1 p_1(t),\\ \frac{\mathrm{d}}{\mathrm{d}t}p_1(t) &= \lambda_0 p_0(t) - (\lambda_1 + \mu_1)p_1(t),\\ p_0(0) &= 1.\end{aligned} \tag{3.54}$$

After applying Laplace transform, one gets the following system of algebraic equations:

$$\begin{aligned}-1 + s\varphi_0(s) &= -\lambda_0\varphi_0(s) + \mu_1\varphi_1(s),\\ s\varphi_1(s) &= \lambda_0\varphi_0(s) - (\lambda_1 + \mu_1)\varphi_1(s).\end{aligned} \tag{3.55}$$

Since for a dubbed system $P^{(0)}(t) = p_0(t) + p_1(t)$, the solution in terms of Cramer's rule has the form

$$\varphi^{(0)}(s) = \varphi_0(s) + \varphi_1(s) = \frac{\begin{vmatrix} 1 & -\mu_1 \\ 0 & \lambda_1 + \mu_1 + s \end{vmatrix} + \begin{vmatrix} \lambda_0 + s & 1 \\ -\lambda_0 & 0 \end{vmatrix}}{\begin{vmatrix} \lambda_0 + s & -\mu_1 \\ -\lambda_0 & \lambda_1 + \mu_1 + s \end{vmatrix}}$$

$$= \frac{s + \lambda_0 + \lambda_1 + \mu_1}{s^2 + s(\lambda_0 + \lambda_1 + \mu_1) + \lambda_0\lambda_1}. \tag{3.56}$$

Here the superscript at $P^{(0)}(t)$ and $\varphi^{(0)}(s)$ denotes that solution has been obtained under the above-mentioned initial conditions.

Using the same procedure as above, one gets in this case

$$P^{(0)}(t) = \frac{1}{s_1^* - s_2^*} \left( s_1^* \exp(s_2^* t) - s_2^* \exp(s_1^* t) \right), \qquad (3.57)$$

where $s_{1,2}^* = -(\alpha^*/2) \pm \sqrt{(\alpha^*/2)^2 - \beta^*}$, $\alpha^* = \lambda_0 + \lambda_1 + \mu_1$, and $\beta^* = \lambda_0 \lambda_1$.

Below we will need a solution of the same system of linear differential equations with the initial condition $p_1(0) = 1$, that is, when the starting moment is a moment of the system's recovery completion. The corresponding system of algebraic equations for Laplace transforms has the form

$$\begin{aligned} s\varphi_0(s) &= -\lambda_0 \varphi_0(s) + \mu_1 \varphi_1(s), \\ -1 + s\varphi_1(s) &= \lambda_0 \varphi_0(s) - (\lambda_1 + \mu_1)\varphi_1(s). \end{aligned} \qquad (3.58)$$

Let us omit routine deductions absolutely similar to those above, and write the result:

$$\varphi^{(1)}(s) = \varphi_0(s) + \varphi_1(s) = \frac{\begin{vmatrix} 0 & -\mu_1 \\ 1 & \lambda_1 + \mu_1 + s \end{vmatrix} + \begin{vmatrix} \lambda_0 + s & 0 \\ -\lambda_0 & 1 \end{vmatrix}}{\begin{vmatrix} \lambda_0 + s & -\mu_1 \\ -\lambda_0 & \lambda_1 + \mu_1 + s \end{vmatrix}}$$

$$= \frac{\lambda_0 + \mu_1}{s^2 + s(\lambda_0 + \lambda_1 + \mu_1) + \lambda_0 \lambda_1}. \qquad (3.59)$$

Inverse Laplace transform is

$$P^{(1)}(t) = \frac{1}{s_1^* - s_2^*} \left[ (s_1^* - \lambda_0 - \lambda_1) \exp(s_2^* t) - (s_2^* - \lambda_0 - \lambda_1) \exp(s_1^* t) \right],$$

$$(3.60)$$

where roots $s_1^*$ and $s_2^*$ are the same as for (3.57). Here again the superscript at $P^{(1)}(t)$ corresponds to the initial condition $p_1(0) = 1$.

### 3.3.5  Stationary Coefficient of Interval Availability

This is one of the most important reliability indices of recoverable systems. For a dubbed system, this index can be written as a formula of total probability:

$$R(t_0) = p_0 P^{(0)}(t) + p_1 P^{(1)}(t), \tag{3.61}$$

where $p_0$ and $p_1$ are stationary probabilities of states "0" and "1", respectively; probabilities $P^{(0)}(t)$ and $P^{(1)}(t)$ are conditional PFFOs for successful completion of an operation, starting at corresponding states. Probabilities $P^{(0)}(t)$ and $P^{(1)}(t)$ are taken from (3.57) and (3.60), respectively. Stationary probabilities $p_0$ and $p_1$ can be found by using (3.49)–(3.51):

$$p_1 = \frac{\mu_2}{\lambda_1} p_2 = \frac{\lambda_0 \mu_2}{\lambda_1 (\lambda_0 \lambda_1 + \lambda_0 \mu_2 + \mu_1 \mu_2)} \tag{3.62}$$

and

$$p_0 = \frac{\mu_1}{\lambda_0} p_1 = \frac{\mu_1 \mu_2}{\lambda_1 (\lambda_0 \lambda_1 + \lambda_0 \mu_2 + \mu_1 \mu_2)}. \tag{3.63}$$

We do not write the obvious final expression for $R(t_0)$ because it is too clumsy. However, approximations for highly reliable systems have rather simple and compact forms, although it is more convenient to write all of them for specific cases (Table 3.3).

**TABLE 3.3   Approximate Formulas of $R(t_0)$ for Four Cases of Highly Reliable Recoverable Dubbed Systems**

| | | | |
|---|---|---|---|
| Loaded redundancy: unrestricted recovery | $\dfrac{1}{1+\gamma^*}\exp\left(-\dfrac{2\gamma\lambda t_0}{1+2\gamma}\right)$, | where $\gamma^* = \dfrac{\gamma^2}{1+2\gamma}$ | and $\gamma = \dfrac{\lambda}{\mu}$ |
| Loaded redundancy: restricted recovery | $\dfrac{1}{1+\gamma^*}\exp\left(-\dfrac{2\gamma\lambda t_0}{1+2\gamma}\right)$, | where $\gamma^* = \dfrac{2\gamma^2}{1+2\gamma}$ | and $\gamma = \dfrac{\lambda}{\mu}$ |
| Unloaded redundancy: unrestricted recovery | $\dfrac{1}{1+\gamma^*}\exp\left(-\dfrac{\gamma\lambda t_0}{1+2\gamma}\right)$, | where $\gamma^* = \dfrac{\gamma^2}{2(1+\gamma)}$ | and $\gamma = \dfrac{\lambda}{\mu}$ |
| Unloaded redundancy: restricted recovery | $\dfrac{1}{1+\gamma^*}\exp\left(-\dfrac{\gamma\lambda t_0}{1+2\gamma}\right)$, | where $\gamma^* = \dfrac{\gamma^2}{1+\gamma}$ | and $\gamma = \dfrac{\lambda}{\mu}$ |

### 3.3.6  Mean Time to Failure

MTTF can be easily found from the standard formula:

$$T^{(0)} = \int_0^\infty P^{(0)}(t)dt. \tag{3.64}$$

Let us note that Laplace transform of function $P_0(t)$ with substitution $s = 0$ also gives $T^{(0)}$. Indeed,

$$\varphi(s)|_{s=0} = \int_0^\infty e^{-st}P^{(0)}(t)dt|_{s=0} = \int_0^\infty P^{(0)}(t)dt = T^{(0)}. \tag{3.65}$$

Thus, $T^{(0)}$ can be found with the help of Laplace transform (3.56):

$$T^{(0)} = \left. \frac{s + \lambda_0 + \lambda_1 + \mu_1}{s^2 + s(\lambda_0 + \lambda_1 + \mu_1) + \lambda_0\lambda_1} \right|_{s=0} = \frac{\lambda_0 + \lambda_1 + \mu_1}{\lambda_0\lambda_1}$$

$$= \frac{1}{\lambda_0} + \frac{1}{\lambda_1} + \frac{\mu_1}{\lambda_0\lambda_1}. \tag{3.66}$$

### 3.3.7  Mean Time Between Failures

For a dubbed system, MTTF and MTBF are different. It is clear because functions $P^{(0)}(t)$ and $P^{(1)}(t)$ are different. MTBF can be found with the help of Laplace transform (3.59):

$$T^{(1)} = \left. \frac{\lambda_0 + \mu_1}{s^2 + s(\lambda_0 + \lambda_1 + \mu_1) + \lambda_0\lambda_1} \right|_{s=0} = \frac{\lambda_0 + \mu_1}{\lambda_0\lambda_1} = \frac{1}{\lambda_0} + \frac{\mu_1}{\lambda_0\lambda_1}. \tag{3.67}$$

Note that $T^{(0)}$ is larger than $T^{(1)}$ by a value of $1/\lambda_0$. It is clear that this is the average "travel" time from state "0" to state "1".

The same result can also be obtained in a different way. From the above statement follows

$$T^{(0)} = \frac{1}{\lambda_0} + T^{(1)}. \tag{3.68}$$

Coming to state "1", the process stays there before going to state "0" or state "2" on average time $1/(\lambda_1 + \mu_1)$ and after that it returns to state "0" with probability $\mu_1/(\lambda_1 + \mu_1)$ or moves to absorbing state "2" with probability $\lambda_1/(\lambda_1 + \mu_1)$. It gives a possibility to write the following recurrent equation:

$$T^{(1)} = \frac{1}{\lambda_1 + \mu_1} + \frac{\mu_1}{\lambda_1 + \mu_1} T^{(0)}. \tag{3.69}$$

Substituting (3.69) into (3.68) gives the final expression:

$$T^{(0)} = \frac{1}{1 - [1/(\lambda_1 + \mu_1)]} \left( \frac{1}{\lambda_0} + \frac{1}{\lambda_1 + \mu_1} \right) = \frac{\lambda_0 + \lambda_1 + \mu_1}{\lambda_0 \lambda_1}$$

$$= \frac{1}{\lambda_0} + \frac{1}{\lambda_1} + \frac{\mu_1}{\lambda_0 \lambda_1}. \tag{3.70}$$

From (3.69) and (3.70) follows that

$$T^{(1)} = T^{(0)} - \frac{1}{\lambda_0} = \frac{1}{\lambda_1} + \frac{\mu_1}{\lambda_0 \lambda_1}. \tag{3.71}$$

We give sometimes several different deductions of the same results exclusively for a single purpose: to help the readers develop "mathematical intuition."

The final results for all types of recoverable dubbed systems are presented in Table 3.4.

**TABLE 3.4    MTTF and MTBF for Four Types of Recoverable Dubbed Systems**

|  | $T^{(0)}$ | | $T^{(1)}$ | |
|---|---|---|---|---|
|  | Exact | Approximate | Exact | Approximate |
| Loaded redundancy | $\dfrac{1}{\lambda} \cdot \dfrac{1 + 3\gamma}{2\gamma}$ | $\dfrac{1}{2\lambda\gamma}$ | $\dfrac{1}{\lambda} \cdot \dfrac{1 + 2\gamma}{2\gamma}$ | $\dfrac{1}{2\lambda\gamma}$ |
| Unloaded redundancy | $\dfrac{1}{\lambda} \cdot \left( 2 + \dfrac{1}{\gamma} \right)$ | $\dfrac{1}{\lambda\gamma}$ | $\dfrac{1}{\lambda} \cdot \left( 1 + \dfrac{1}{\gamma} \right)$ | $\dfrac{1}{\lambda\gamma}$ |

### 3.3.8  Mean Recovery Time

The mean recovery time, $\tau$, for this simple Markov model coincides with average duration the process stays in state "2". As follows from transition graphs, this time is equal to $1/\mu_2$. So, this reliability index depends on the number of repair facilities. For unrestricted recovery (two repair facilities), the mean recovery time is $1/2\mu$, and for restricted repair facilities (a single repair facility), $1/\mu$.

## 3.4  PARALLEL SYSTEMS

By definition, a parallel system is factually a single operating unit with a group of identical redundant units, which are independent in sense of failing. Such an idealized scheme has few relations to a real engineering practice, although it is of theoretical interest. Speaking about parallel systems consisting of recoverable units, one has to keep in mind four main possible cases, presented in Table 3.5.

Probably, the mathematical description of such systems is given by "birth–death process" that is considered in detail in Appendix C.2.

Let a parallel system consists of $n$ units, that is, one operating unit and $n-1$ redundant ones. Assume that there are $k$ repair facilities for recovering failed units, $k \le n - 1$. Let us denote states by natural

**TABLE 3.5  Main Cases of Parallel Systems**

| | | Regime of Recovering | |
|---|---|---|---|
| | | Unrestricted Repair | Restricted Repair |
| Regime of redundant units | Loaded | Factually, all $n$ units in parallel are independent | As soon as the number of failed units exceeds the number of repair facilities, failed units form a waiting line |
| | Unloaded | Failure of such a system occurs only if during repair of first failed units all other units have failed | As in above case, when the number of failed units exceeds the number of repair facilities, failed units form a waiting line |

numbers 0, 1, 2, . . . , where the number of a state corresponds to the number of failed units. Then all four cases can be described, actually, by very similar linear transition graphs (Figure 3.10).

For each of these transition graphs, the system of differential equations, corresponding system of algebraic equations for Laplace transforms, and algebraic equations for stationary probabilities can be easily found with the help of Appendix B. We will omit them, first of all, because "pure parallel" systems with multiple loaded redundant units are rare in engineering practice. Much more interesting is the structure that is described in the next section.

## 3.5 STRUCTURES OF TYPE "$M$-OUT-OF-$N$"

Much more realistic is a series system of identical independent units with a common group of redundant units. Formally, such a structure appears if a system consists of units of several types. A set of units of the same type can be considered as a "series system," for which there is a stock of spare units. It is reasonable to consider these spare units as unloaded; these units wait for being switched into operating position after one of operating units has failed. Failed units are directed to a repair shop, from where after recovery they again enter the system's stock. Switching of spare unit into an operating position is usually assumed instantaneous. (Of course, this assumption is almost correct if switching time is relatively small.)

In this case, the system as a whole can be presented as a series connection of such "$m_k$-out-of-$n_k$" subsystems (Figure 3.12).

Transition graph for one of such series subsystems of $m$ operating units and common group of $(n - m)$ unloaded spare units is presented in Figure 3.13). States of system failure are shadowed.

This model of redundancy is, probably, one of the most useful models for practical purposes in reliability engineering. We will repeat most of deductions given in Appendix C.2 for this particular, however very important, case. We omit only technical details minutely described in the mentioned appendix.

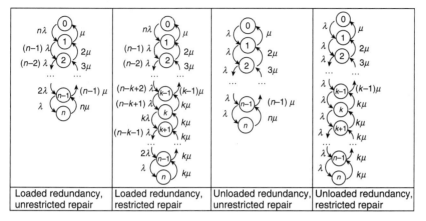

| Loaded redundancy, unrestricted repair | Loaded redundancy, restricted repair | Unloaded redundancy, unrestricted repair | Unloaded redundancy, restricted repair |
|---|---|---|---|

**FIGURE 3.11** Transition graphs for corresponding "birth–death processes."

The system of differential equations in this case takes the form

$$\frac{d}{dt}p_0(t) = -m\lambda p_0(t) + \mu p_1(t),$$

$$\frac{d}{dt}p_1(t) = m\lambda p_0(t) - p_1(t)(m\lambda + 2\mu) + 2\mu p_2(t),$$

$$\vdots$$

$$\frac{d}{dt}p_{n-m-1}(t) = m\lambda p_{n-m-2}(t) - [m\lambda + (n-m-1)\mu]p_{n-m-1}(t)$$
$$+ (n-m)\mu p_{n-m}(t),$$

$$\frac{d}{dt}p_{n-m}(t) = m\lambda p_{n-m-1}(t) - [m\lambda + (n-m)\mu]p_{n-m}(t)$$
$$+ (n-m+1)\mu p_{n-m+1}(t),$$

$$\frac{d}{dt}p_{n-m+1}(t) = m\lambda p_{n-m}(t) - [(m-1)\lambda + (n-m+1)\mu]p_{n-m+1}(t)$$
$$+ (n-m+2)\mu p_{n-m+2}(t),$$

$$\vdots$$

$$\frac{d}{dt}p_{n-1}(t) = 2\lambda p_{n-2}(t) - [\lambda + (n-1)\mu]p_{n-1}(t) + n\mu p_n(t),$$

$$\frac{d}{dt}p_n(t) = \lambda p_{n-1}(t) - n\mu p_n(t).$$

$$(3.72)$$

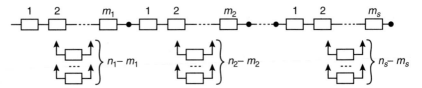

**FIGURE 3.12**   Conditional presentation of a system as a series connection of "*m*-out-of-*n*" subsystems.

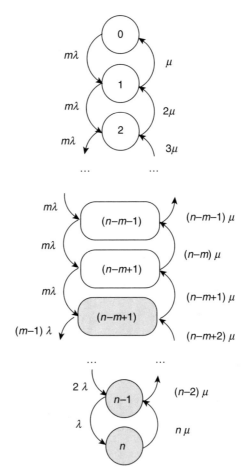

**FIGURE 3.13**   Transition graph of an "*m*-out-of-*n*" system with unloaded spare units.

The initial condition in most reliability applications is taken in the form $p_0(0) = 1$.

Of course, system of equations (3.72) can be solved with the help of methods described in Appendix NaN. The probability of failure-free operation can be found as

$$P(t) = \sum_{0 \leq i \leq n-m} p_i(t). \tag{3.73}$$

We will not spend time for pure mathematical exercises because such systems are oriented on long run, and, consequently, stationary availability coefficient for such systems is more appropriate reliability index rather than PFFO. In this case, the system (3.72) transforms into the system of algebraic equations:

$$m\lambda p_0 = \mu p_1,$$
$$(m\lambda + 2\mu)p_1 = m\lambda p_0 + 2\mu p_2,$$
$$\vdots$$
$$[m\lambda + (n - m - 1)\mu]p_{n-m-1} = m\lambda p_{n-m-2} + (n - m)\mu p_{n-m},$$
$$[m\lambda + (n - m)\mu]p_{n-m} = m\lambda p_{n-m-1} + (n - m + 1)\mu p_{n-m+1},$$
$$[(m - 1)\lambda + (n - m + 1)\mu]p_{n-m+1} = m\lambda p_{n-m} + (n - m + 2)\mu p_{n-m+2},$$
$$\vdots$$
$$[\lambda + (n - 1)\mu]p_{n-1} = 2\lambda p_{n-2} + n\mu p_n,$$
$$\lambda p_{n-1} = n\mu p_n. \tag{3.74}$$

Since equations in (3.74) are mutually dependent, it is necessary additionally to use equation of total probability:

$$\sum_{k=0}^{n} p_k = 1. \tag{3.75}$$

As stated in Appendix C.2, actually it is more convenient to write the equations of balance for "cuts" of the transition graph rather than for states. It actually leads to the solution almost directly. Remember that the balance means that flows back and forth through a cut between neighbor states of the transition graph are equal. Thus, on the basis of the transition graph depicted in Figure 3.13, we can write

$$
\begin{aligned}
m\lambda p_0 &= \mu p_1, \\
m\lambda p_1 &= 2\mu p_2, \\
m\lambda p_2 &= 3\mu p_3,
\end{aligned}
$$

$$\vdots$$

$$
\begin{aligned}
m\lambda p_{n-m-1} &= (n-m)\mu p_{n-m}, \\
m\lambda p_{n-m} &= (n-m+1)\mu p_{n-m+1}, \\
(m-1)\lambda p_{n-m+1} &= (n-m+2)\mu p_{n-m+2},
\end{aligned}
\tag{3.76}
$$

$$\vdots$$

$$
\begin{aligned}
(n-1)\mu p_{n-1} &= 2\lambda p_{n-2}, \\
\lambda p_{n-1} &= n\mu p_n.
\end{aligned}
$$

Introducing, for convenience, notation $\rho = \lambda/\mu$, one gets system (3.76) in the form

$$
\begin{aligned}
p_1 &= m\rho p_0, \\
p_2 &= \frac{m}{2}\rho p_1 = \frac{m^2}{1\cdot 2}\rho^2 p_0, \\
p_3 &= \frac{m}{3}\rho p_2 = \frac{m^3}{1\cdot 2\cdot 3}\rho^3 p_0,
\end{aligned}
$$

$$\vdots$$

$$
\begin{aligned}
p_{n-m} &= \frac{m}{n-m}\rho p_{n-m-1} = \frac{m^{n-m}}{(n-m)!}\rho^{n-m}p_0, \\
p_{n-m+1} &= \frac{m}{n-m+1}\rho p_{n-m} = \frac{m\cdot m^{n-m}}{(n-m+1)!}\rho^{n-m+1}p_0, \\
p_{n-m+2} &= \frac{m-1}{n-m+2}\rho p_{n-m+1} = \frac{m\cdot(m-1)\cdot m^{n-m}}{(n-m+2)!}\rho^{n-m+2}p_0, \\
p_{n-m+3} &= \frac{m-2}{n-m+3}\rho p_{n-m+2} = \frac{m\cdot(m-1)\cdot(m-2)\cdot m^{n-m}}{(n-m+3)!}\rho^{n-m+3} \\
p_0 &= \frac{m!}{(m-3)!}\cdot\frac{m^{n-m}}{(n-m+3)!}\rho^{n-m+3}p_0,
\end{aligned}
$$

$$\vdots$$

$$
\begin{aligned}
p_{n-2} &= \frac{n-2}{3}\rho p_{n-1} = \frac{m!}{3!}\cdot\frac{m^{n-m}}{(n-2)!}\rho^{n-2}p_0, \\
p_{n-1} &= \frac{n-1}{2}\rho p_{n-1} = \frac{m!}{2!}\cdot\frac{m^{n-m}}{(n-1)!}\rho^{n-1}p_0, \\
p_n &= \frac{n}{1}\rho p_{n-1} = \frac{m!}{1!}\cdot\frac{m^{n-m}}{n!}\rho^n p_0.
\end{aligned}
\tag{3.77}
$$

Using (3.75), one can write the solution

$$p_0 = \left(1 + \sum_{k=1}^{n-m} \frac{m^k}{k!}\rho^k + \sum_{k=n-m+1}^{n} \frac{m^{n-m}}{k!}\rho^k \prod_{k=n-m+1}^{n-k} \frac{(m-1)!}{(k-n+m-1)!}\right)^{-1}.$$

(3.78)

For any $p_k$, solution can be easily obtained by substitution of $p_0$ in the corresponding equation of (3.77).

# 4

# RECOVERABLE SYSTEMS: HEURISTIC MODELS

## 4.1 PRELIMINARY NOTES

In the previous chapter, we presented analysis of recoverable parallel systems by the means of Markov models. The reader had a chance to receive evidence that even idealized simplest models need rather sophisticated mathematical technique. What if the model is a little bit more realistic? For instance, the dubbed system has a nonreliable switch? What if an operating unit (unit on operational position) does not have complete monitoring of its state, and hence can lead to a hidden system failure? (Of course, such kinds of important practical features of redundant systems can be continued.) In Figure 4.1, a transition graph for the system described above is presented with an additional condition: repair failed part is performing in accordance with the rule FIFO (first-in, first-out). In this case, the transition graph is rather ugly!

*Probabilistic Reliability Models*, First Edition. Igor Ushakov.
© 2012 John Wiley & Sons, Inc. Published 2012 by John Wiley & Sons, Inc.

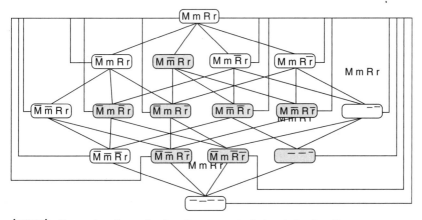

Legend:  **M** - monitored part of main unit; m - uncontrolled part of main unit;
**R** - monitored part of redundant unit; r - uncontrolled part of redundant unit;
‒ means failure of corresponding part;
——— means recovering transition;
⟍ means transition due to failure;
▭ – operartional system state;
▭ – state of system failure (probably, hidden).

**FIGURE 4.1**    Transition graph for a "simple" recoverable dubbed system with unreliable switching and incomplete control of operating and redundant units.

Hardly anyone will take the courage to solve a corresponding system of differential or even algebraic equations!

At the same time, such a problem can be easily solved with the help of heuristic methods.

Do not think that heuristic is something taken "from the ceiling." The heuristic we are talking about is an approximation method based on strong mathematical asymptotical results working for analysis of highly reliable systems.

However, what to do if the system is not highly reliable? The answer is simple: in this case, you should think of improving system reliability, and not spending your time on senseless "reliability analysis," deducting ugly and useless "five-store" formulas.

**FIGURE 4.2**   "Columbus breaking the egg" by William Hogarth (1697–1764).

A researcher is often faced with the problem of finding a "solution" to a problem when the problem is practically unsolvable. However, in spite of the problem's "insolvability," a solution must be found! And even if an exact or "ideal" solution cannot be found, a designer is forced to make a practical decision since the required problem has to be solved!

Remember that famous legend about "Columbus's egg" (Figure 4.2)? It refers to a brilliant idea or discovery that seems simple and easy after the fact. That story of how Christopher Columbus, having been told that discovering the Americas was no great accomplishment, challenged his critics to make an egg stand on its tip. After his challengers gave up, Columbus did it himself: he tapped it gently on the table breaking it slightly, and with this the egg stood on its end because he had flattened its tip. His sailors were buzzing: "We would do the same!" Columbus answered: "You would; however, I did it!"

When there is no exact analytical solution and the problem is still too hard for even a Monte Carlo simulation, the only possibility is to use a heuristic procedure (heuristics).

Sometimes heuristics are thought to lead to an arbitrary "solution," based only on a "I personally believe" type of argument. We oppose such "heuristics," as we understand the term "heuristic" to be an extension of analytical methods in areas where such methods cannot be exactly proven. Sometimes we omit some specified conditions, and sometimes we make additional assumptions and are not sure that the method of solution is still correct. Sometimes we change an analyzed phenomenon description to allow the use of available mathematical tools.

In fact, the building of a mathematical model is always a heuristic procedure itself. No mathematical model completely reflects all of the properties of a real object. We always create "an ideal image" based on a real object, and then build a mathematical model for this idealized image.

Moreover, approximate calculations can be viewed as "good proven heuristic."

Thus, heuristics is an inevitable part of mathematical modeling.

Below we will introduce several heuristic approaches. Generally, they concern the construction of models.

Anyway, never use a cannon for hunting birds: a simple fowling piece is quite enough for this purpose.

## 4.2   POISSON PROCESS

In reliability theory, the Poisson process occupies a special place. Models based on the Poisson process are rather simple and very convenient for numerical analysis. However, it is not like the drunk who is looking for his lost keys under the street lamp because that is the most illuminated place, not because that is the place where he dropped the keys. The utilization of the Poisson models in reliability has a very strong empirical background.

Most of the complex systems consist of a large number of relatively reliable units. Flow of system failures is generated by many "subflows" of units' failures. These "subflows" are mutually independent because they are formed by independent units; for highly reliable units probability of intersection of failures is negligible and, finally, after relatively short time the process of system's failure becomes stationary, that is, it does not change its probabilistic properties with time, and these properties become very close to the properties of the Poisson process: the Markov property, the orderliness, and stationarity. Let us explain these properties in more detail.

1. *Markov property* means that the future development of the process does not depend either on current state of the process or on its entire prehistory.

2. *Orderliness* means that there are no simultaneously happening events or even the so-called "points of concentration." In other words, if time interval $\Delta \to 0$, the probability of occurrence of more than one event, $p_{k>1}(\Delta)$, within this interval becomes infinitesimally small in comparison to $p_1(\Delta)$:

$$\lim_{\Delta \to 0} \frac{p_{k>1}(\Delta)}{p_1(\Delta)} \to 0.$$

3. *Stationarity* means invariance to the shift operator; that is, probabilistic characteristics of the process (for instance, the mean number of failures in an interval of given length $t$) depends on the length of the interval and does not depend on its location on time axis.

The Poisson process especially well describes the process of generating electronic equipment failures. Not in vain, the Poisson process is often called "the process of rare events." In the theory of stochastic processes, the Poisson process plays a role that is analogous to that of the normal distribution in probability theory.

The Poisson process is defined as a point stochastic process that is formed by a sequence of independent random variables with the same exponential distribution. Probability that within a fixed time interval $[x, x + t]$ exactly $k$ events of the Poisson process occur has the Poisson distribution:

$$p_k(x, x + t) = \frac{(\lambda t)^k}{k!} \exp(-\lambda t). \tag{4.1}$$

Parameter $\lambda$ is called *intensity of the Poisson process*. By definition, it is equal to the mean number of events (in our case, failures) in a unit of time, that is, $\lambda = 1/T$, where $T$ is average distance between events (in our case, $T$ is MTBF).

Actually, formula (4.1) complies with the three properties that were formulated almost on qualitative level. Indeed, take an arbitrary fixed interval of length $t$ and divide it into small subintervals $\Delta_i$ so that $\sum_{\forall i} \Delta_i = t$.

The mean number of failures within interval $\Delta_i$ is

$$E\{\Delta_i\} = \lambda \Delta_i. \tag{4.2}$$

On the other hand, for the same interval $\Delta_i$, one can write another expression for the mean number of failures:

$$E\{\Delta_i\} = \sum_{0 \leq j < \infty} j p_j(\Delta_i), \tag{4.3}$$

where $p_k(\Delta_i)$ is the probability that exactly $k$ failures have occurred within interval $\Delta_i$. Taking into account the property of orderliness, the following equation can be written for total probability:

$$p_0(\Delta_k) + p_1(\Delta_k) + o(\Delta_k) = 1, \tag{4.4}$$

where $o(\Delta_k)$ means infinitesimally small value in comparison to $p_1(\Delta_k)$. Thus, property of orderliness allows us to rewrite (4.3) as follows:

$$\lim_{\Delta_k \to 0} E(\Delta_k) = p_1(\Delta_k) \tag{4.5}$$

or, taking into account (4.2) and (4.4),

$$p_1(\Delta_i) = \lambda \Delta_i \tag{4.6}$$

and

$$p_0(\Delta_i) = 1 - \lambda \Delta_i. \tag{4.7}$$

Now let us find the probability of no failures within entire interval of length $t$. Due to the Markov property,

$$P(t) = \prod_{\forall i} p_0(\Delta_i). \tag{4.8}$$

In limit, under condition that all $\Delta_i$ tend to 0, and remembering that $\sum \Delta_i = t$, we get

$$P(t) = \lim_{\Delta \to 0} (1 - \lambda\Delta)^{t/\Delta} = \lim_{\Delta \to 0} (1 - \lambda\Delta)^{\frac{1}{(-\lambda\Delta)(-\lambda t)}} = \exp(-\lambda t).$$

(4.9)

This completes the proof that the three characterization properties of the Poisson process lead to recurrent point process with intervals between events distributed exponentially.

## 4.3   PROCEDURES OVER POISSON PROCESSES

Before beginning with explanations of a suggested heuristic method, let us consider two simple procedures: *thinning* of the Poisson process and *superposition* of the Poisson processes.

### 4.3.1   Thinning Procedure

Procedure of thinning consists in the following. Let us exclude from an arbitrary Poisson process points with constant probability $q$ and keep them with probability $p = 1 - q$. What kind of process we will get in this case?

Get the result without any special mathematical proof based only on characterization properties of the Poisson process.

Markov property of a new point process conserves because probability of exclusion of any point does not depend on entire prehistory of the process.

Orderliness property conserves due to the fact that points are only excluded from the process.

Stationarity of the process conserves due to the constant value of probability $q$ on all time axes (Figure 4.3).

Thus, since all characterization properties have been preserved under such a thinning procedure, the resulting process is the Poisson one.

Of course, not any "thinning procedure" with the Poisson process leads to such results. Assume that probability of point exclusion, for instance, is decreasing in time. In this case, the flow of failures will

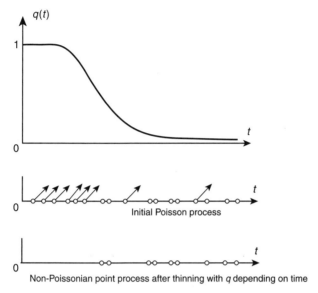

FIGURE 4.3    Example of thinning with probability $q$ depending on time.

remind of a traditional aging process of a system consisting of wearing out units: before "complete death" the system will fail more and more often; that is, no properties of the Poisson process are observed.

Another example of thinning that does not conserve the Markov property is thinning with probabilities of exclusion points from the process depending on the ordinal number of points in the initial process. For instance, let $q_k = 1$ for all $k \neq 2^{3x}$ and $k \neq 2^{3x+1}$, where $x$ is

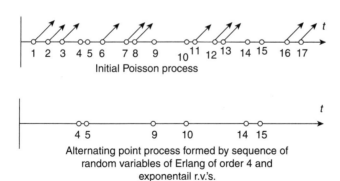

**FIGURE 4.4**    Example of thinning with probability $q$ depending on ordinal number of points in the initial process.

a natural number, and $q_k = 0$, otherwise. In other words, we have a new point process: after interval presenting sum of three i.i.d. exponential random variables with probability 1 follows a single exponentially distributed time interval. In other words, the process becomes an alternative process formed by an alternative sequence of intervals with Erlang and exponential distribution (see Figure 4.4).

### 4.3.2 Superposition Procedure

The second important procedure is the so-called superposition of independent Poisson processes.

Markov property of a new point process conserves due to independence of initial processes and the Markov property of each ingredient process.

Orderliness property conserves due to the fact that exponential distribution is continuous; hence, exact coincidence of two independent random variables is impossible. Due to the same reason, appearance of points of concentrations also becomes an impossible event.

Stationarity of the resulting process conserves due to initial stationarity of each incoming ingredient.

Thus, since all characterization properties of the Poisson process have been preserved under such a procedure of superposition, the resulting process is also the Poisson one with parameter equal to sum of parameters of inputting subprocesses.

These simple explanations are given only to provide the reader more exact filling of these two procedures that are so important in theory of stochastic point processes.

The approximate methods suggested below are based on two important limit theorems in the theory of point stochastic processes.

## 4.4   ASYMPTOTIC THINNING PROCEDURE OVER STOCHASTIC POINT PROCESS

We begin with a renewal process. By definition, a counting process for which the interarrival times are i.i.d. with an arbitrary so-called

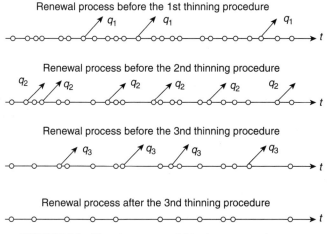

Renewal process before the 1st thinning procedure

Renewal process before the 2nd thinning procedure

Renewal process before the 3nd thinning procedure

Renewal process after the 3nd thinning procedure

**FIGURE 4.5**   First three steps of thinning a renewal process.

"forming distribution" $F(t)$ is said to be a renewal process. In this sense, the Poisson process is a particular case of a renewal one.

Let us apply the thinning procedure to this process multiply, excluding points step by step as shown in Figure 4.5.

The strong definition of such an asymptotic thinning procedure is known as the Renyi[1] limit theorem.

If $Q_n = q_1 q_2 \cdots q_n \to 0$ as $n \to 8$, then in limit the thinned renewal process approaches the Poisson one. Of course, such a procedure leads to infinite growth of the interarrival times. To keep the resulting process of the same intensity (i.e., with the same average length of intervals as in the initial renewal process), one needs simultaneously to "compress" the time axis: new timescale after $n$ thinning procedures has to be $Q_n t$.

Some time later, Yu. Belyaev[2] generalized the result on stochastic point processes beyond renewal ones.

---

[1] Alfred Renyi (1921–1970) was a Hungarian mathematician who was known for his contributions in combinatorics, graph theory, number theory, but mostly in probability theory.

[2] Yuri Konstantinovich Belyaev (b. 1932) is a Russian statistician, Professor of Moscow State University, and Professor Emeritus of University of Umeå (Sweden). He is a pupil of A. N. Kolmogorov.

## 4.5   ASYMPTOTIC SUPERPOSITION OF STOCHASTIC POINT PROCESSES

Another important asymptotic result is the Khinchin[3]–Ososkov[4] theorem of the superposition (union) of independent renewal processes (Figure 4.6). Later, this result was independently expanded on more general stochastic point processes by B. Grigelionis[5] and I. Pogozhev[6] in the theorem named after them.

The theorem states that superposition of $n$ independent renewal processes forms in limit the Poisson process, when $n \to \infty$. Indeed, let us again check if such a procedure leads to three properties of the Poisson process.

Each renewal process, including the resulting one, possesses the property of "restricted aftereffect"; that is, its future depends only on the moment of the last occurred event. If the number of superposed renewal processes is large, the process, in a sense, "forgets" its past: too many other independent events "intervene" between two neighbor

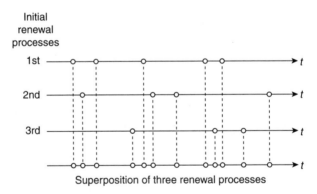

Superposition of three renewal processes

**FIGURE 4.6**   Example of superposition of three renewal processes.

---

[3] Alexander Yakovlevich Khinchin (1894–1959) was a Soviet mathematician and one of the most significant people in the Soviet school of probability theory.
[4] Gennady Alexeevich Ososkov (b. 1931) is a Soviet and Russian mathematician. He is a pupil of A. N. Kolmogorov and A. Ya. Khinchin.
[5] Bronyus Igno Grigelionis (b. 1935) is a Lithuanian mathematician. He is a pupil of B. V. Gnedenko.
[6] Ivan Borisovich Pogozhev (1923–2011) was a Soviet and Russian mathematician.

arrivals of the same process. Thus, it is understandable, on intuitive level, that process asymptotically begins to possess the Markov property.

Orderliness property conserves due to the fact that "forming distributions" $F_k(t)$ of initial renewal processes are continuous, and due to it exact coincidence of two arrivals or appearance of points of concentrations are impossible events.

Stationarity of the resulting process is delivered by the fact that each of the initial renewal processes has its own constant intensity.

Thus, since all characterization properties of the Poisson process are presented, the resulting point process asymptotically approaches the Poisson one with parameter equal to sum of parameters of inputting subprocesses.

Of course, in this case interarrival intervals go to 0, so we again should change timescale: we have to "stretch out" the time axes to keep length of interarrival intervals in a reasonable scale.

The main requirement for correctly using the theorem of superposition is the condition of "uniformity" of point processes that compose the resulting process. For instance, assume that we superpose a single regular point process with constant interarrival time, $\tau$, and infinite number of Poisson processes of such type that their intensity, $\lambda_k$, decreases in such a way that $\lim_{n \to \infty} \sum_{1 \le k \le n} \lambda_k = \Lambda$, where $\Lambda \ll 1/\tau$ (see Figure 4.7).

From this figure, one can see that points of the regular process prevail over points of other processes.

Now we will show that both of these theorems are extremely constructive and effective for a heuristic analysis of highly reliable repairable systems. Of course, rigid asymptotic results used in a prelimit case give only approximations; however, what does it mean by a "correct model"? A model is always a model. We would like to underline once more that model construction is "an art," and preciseness of probabilistic analysis of real objects and phenomena depends, in first turn, on their understanding, rather than on filigree mathematical exercises.

Everybody understands that the main imperfection of any heuristic method usually lies in the impossibility of defining the domain where

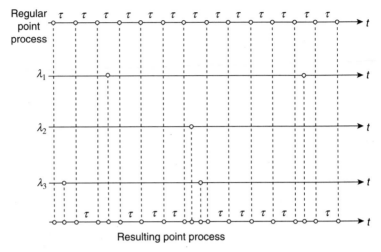

**FIGURE 4.7** Superposition of a "dense" regular point process with a number of "weak" Poisson processes.

the results obtained with its help are valid. But in this particular case, we can suggest a simple (and convenient!) rule: if you have obtained a high value of a reliability index, the application of the heuristic method *was correct*. It is not a bad rule because otherwise a system is improper for practical use!

The best way to explain a heuristic method in detail is to show examples of how it works.

## 4.6   INTERSECTION OF FLOWS OF NARROW IMPULSES

The main idea is provided in the following. Consider two alternating renewal processes (flows of impulses), one with interarrival time $T_1$ and impulse width $\tau_1$, and another with corresponding parameters $T_2$ and $\tau_2$. In assumption of exponentiality of interarrival time, one can write $\lambda_1 = 1/T_1$ and $\lambda_2 = 1/T_2$.

Intersection of two impulses can happen if the front edge of an impulse of the first flow appears within an impulse of the second flow, or if the front edge of an impulse of the second flow appears within an impulse of the first flow (see Figure 4.8).

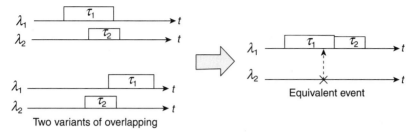

**FIGURE 4.8**   Demonstration of equivalency of two models of impulse overlapping.

Now let us apply these results and above-mentioned asymptotic theorems to reliability analysis of simplest series and parallel systems.

First, consider a series system of two units. What happens if two intervals of downtime overlap? Consider the case of restricted recovery (the unit failed during recovery of the previously failed one is waiting in line). The total recovery time is $\tau_\Sigma = \tau_1 + \tau_2$ (Figure 4.9).

As a result, there are three ingredients of the resulting flow:

1. Flow of impulses of the mean width equals $\tau_\Sigma = \tau_1 + \tau_2$. Intensity of this flow equals

$$\Lambda \approx \lambda_1(\lambda_2\tau_1) + \lambda_2(\lambda_1\tau_2) = \lambda_1\lambda_2(\tau_1 + \tau_2). \qquad (4.10)$$

2. Flow of impulses with the mean width $\tau_1$. Intensity of this flow equals $\lambda_1^* \approx \lambda_1 - \lambda_1\lambda_2\tau_1$.
3. Flow of impulses with the mean width $\tau_2$. Intensity of this flow equals $\lambda_2^* \approx \lambda_2 - \lambda_2\lambda_1\tau_2$.

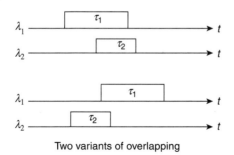

Two variants of overlapping

**FIGURE 4.9**   Explanation of forming the total recovery time in case of restricted recovery.

Being practical, let us evaluate possible values of the found parameters.

If a system is highly reliable, for instance, its availability coefficient is at least 0.95 and the number of units is about 50, then availability coefficient of each unit has to be of order $K = \sqrt[50]{0.95} \approx 0.999$. Assuming that mean downtime is about 0.5 h, it leads to a failure rate of about $0.002\,h^{-1}$. In such assumptions, $\Lambda \approx 0.002^2\,h^{-1} = 0.000004\,h^{-1}$, that is, in practice no sense in "catching the fleas."

Thus, considering series systems with such minor corrections has no sense.

Now consider a dubbed system keeping the same notations for units' parameters. In this case, the system failure occurs if and only if two impulses of downtime overlap. It means that the system failure rate is approximately $\Lambda$, which is defined in (4.10). In this case, the total recovery time is equal to the duration of time interval when both units have failed. This event may occur in two ways, as depicted in Figure 4.10. If recovering time is exponentially distributed, the total system recovery intensity is equal to

$$\mu = \frac{1}{\tau_1} + \frac{1}{\tau_2} = \frac{\tau_1 + \tau_2}{\tau_1 \tau_2} \tag{4.11}$$

or

$$\tau_\Sigma = \frac{\tau_1 \tau_2}{\tau_1 + \tau_2}. \tag{4.12}$$

In Gnedenko and Ushakov (1994), it is shown that (4.11) is valid, at least approximately, for a wide class of distributions.

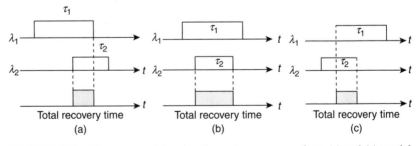

**FIGURE 4.10**    Three ways of forming the system recovery time: (a) and (c) partial overlapping and (b) complete overlapping.

## 4.7 HEURISTIC METHOD FOR RELIABILITY ANALYSIS OF SERIES RECOVERABLE SYSTEMS

Let a series system consist of $n$ units. The $i$th unit has a MTTF $T_i$ and a mean repair time $\tau_i$. In accordance with theorem about asymptotic behavior of point process superposition, failure rate of such a series system is

$$\lambda_{\text{Syst.}} = \left( \sum_{1 \leq i \leq n} \frac{1}{T_i} \right). \tag{4.13}$$

The system's mean downtime can be calculated as a weighted average:

$$\tau_{\text{Syst.}} = \frac{1}{\lambda_{\text{Syst.}}} \sum_{1 \leq i \leq n} \lambda_i \tau_i. \tag{4.14}$$

The system's MDT has hyperexponential distribution (see Appendix NaN).

Probability of failure-free operation is

$$P(t_0) \approx \exp\left( -t_0 \sum_{i=1}^{n} \frac{1}{T_i} \right). \tag{4.15}$$

Availability coefficient is

$$K = \frac{1}{1 + \lambda_{\text{Syst.}} \tau_{\text{Syst.}}} \approx 1 - \sum_{i=1}^{n} \lambda_i \tau_i. \tag{4.16}$$

Operational availability coefficient is

$$R(t_0) \approx \left( 1 - \sum_{i=1}^{n} \lambda_i \tau_i \right) \cdot \left( 1 - t_0 \sum_{i=1}^{n} \lambda_i \right) \approx \left( 1 - \sum_{i=1}^{n} \lambda_i (\tau_i + t_0) \right). \tag{4.17}$$

## 4.8 HEURISTIC METHOD FOR RELIABILITY ANALYSIS OF PARALLEL RECOVERABLE SYSTEMS

Presenting use of the heuristic method to recoverable parallel systems, we restricted ourselves to dubbed systems.

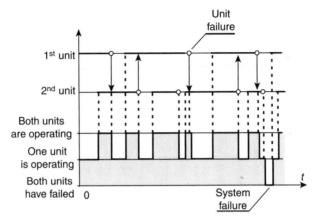

**FIGURE 4.11**    Operating process of a dubbed recoverable system.

The operating process of dubbed recoverable systems can be illustrated by the time diagram shown in Figure 4.11.

Let us begin with a stationary availability coefficient. For a single recoverable unit, this coefficient has the form $1/[1 + (\tau/T)]$, where $\tau$ is MDT and $T$ is MTBF. If time between failures has exponential d.f., one can write the approximation $1 - \lambda\tau$.

Thus, for a dubbed system availability coefficient is approximately equal to

$$K \approx 1 - [1 - (1 - \lambda\tau)]^2 = 1 - (\lambda\tau)^2. \qquad (4.18)$$

Now we consider some important special cases.

### 4.8.1  Influence of Unreliable Switching Procedure

In reality, a dubbed system can fail during the process of switching from a failed operating unit to a redundant one.

If $\pi$ is the probability of successful switching, then the system will fail, on average, after $1/(1 - \pi)$ switching. In other words, dubbed system failure rate due to this particular cause is equal to $\lambda(1 - \pi)$. Assume that MDT in case of switching process failure is $\tau_s$. Another ingredient of system's failure flow is simultaneous failure of both units. The failure rate due to this case is $\lambda^2\tau$ and MDT is $0.5\tau$. These simple considerations lead to the RBD shown in Figure 4.12.

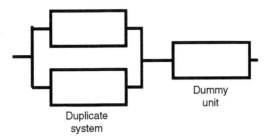

Duplicate
system

Dummy
unit

**FIGURE 4.12**  Conditional RBD of a dubbed system with an unreliable switching procedure.

**TABLE 4.1  Dependence of Availability Coefficient on Reliability of Switching Procedure**

| $\pi$ | Unit's Coefficient of Availability | | |
|---|---|---|---|
| | 0.9 | 0.95 | 0.99 |
| 0.9 | 0.979 | 0.992 | 0.9989 |
| 0.95 | 0.985 | 0.995 | 0.9994 |
| 0.99 | 0.989 | 0.997 | 0.9998 |
| 1 | 0.99 | 0.998 | 0.9999 |

In this case, the resulting availability coefficient of a dubbed system can be written as

$$K \approx \left[1 - (\lambda \tau)^2\right] \cdot \frac{1}{1 + \lambda(1 - \pi)\tau_s} \approx 1 - (\lambda \tau)^2 - \lambda \tau_s (1 - \pi). \quad (4.19)$$

Dependence of availability coefficient on the level of switching reliability is presented in Table 4.1.

## 4.8.2 Influence of Switch's Unreliability

Sometimes a switch plays the role of a distinctive interface. It means that the switch failure leads to the system failure. The RBD for this case is presented in Figure 4.13.

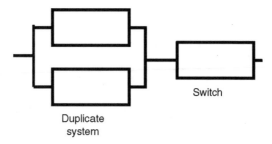

FIGURE 4.13    RBD for a dubbed system and a switch interface.

TABLE 4.2    Dependence of Availability Coefficient on Availability of Switch
Interface

| $K_s$ | Unit's Availability Coefficient | | |
|---|---|---|---|
| | 0.9 | 0.95 | 0.99 |
| 0.9 | 0.891 | 0.95 | 0.99 |
| 0.95 | 0.94 | 0.898 | 0.8999 |
| 0.99 | 0.98 | 0.948 | 0.9499 |
| 1 | 0.99 | 0.9975 | 0.9999 |

In this case, the system availability coefficient can be written as

$$K \approx [1 - (\lambda\tau)^2] \cdot K_s, \tag{4.20}$$

where $K_s$ is the availability coefficient of the switch interface.

In Table 4.2, one can find some numerical illustrations.

### 4.8.3    Periodical Monitoring of the Operating Unit

Sometimes for checking if an operating unit of a dubbed system is
operational or not, one has to perform special periodical testing. In this
situation, a system fails after operating unit failure until it will have
been detected. Thus, the system is found in the state of undetected failure during time interval with average length $0.5\theta$, where $\theta$ is the period
of testing (see Figure 4.14).

In this case, system's availability coefficient can be written as

$$K \approx [1 - (\lambda\tau)^2] \cdot (1 - 0.5\lambda\theta) \approx 1 - (\lambda\tau)^2 - 0.5\lambda\theta. \tag{4.21}$$

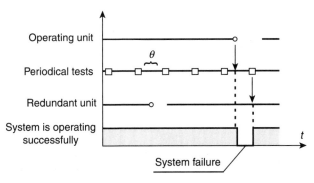

**FIGURE 4.14**   Time diagram illustrating the formation of system undetected failure.


**TABLE 4.3   Dependence of Availability Coefficient on Frequency of Operating Unit Testing**

|  | Unit's Availability Coefficient | | |
| --- | --- | --- | --- |
| $\theta/T$ | 0.9 | 0.95 | 0.99 |
| 0.1 | 0.94 | 0.948 | 0.95 |
| 0.02 | 0.98 | 0.988 | 0.99 |
| 0.01 | 0.98 | 0.993 | 0.995 |
| 0 | 0.99 | 0.9975 | 0.9999 |

Formula (4.21) shows that the testing period has to be significantly less than a single unit MTBF, otherwise there will be null effect of redundancy. Numerical examples are given in Table 4.3.


### 4.8.4   Partial Monitoring of the Operating Unit

Efficiency of a dubbed system depends on completeness of monitoring of an operating unit, which is currently performing the needed operation. Usually, testing of a redundant unit is made easier than that of a unit on operational position because testing itself can interfere with operating functions of the system. Consider a dubbed system with a partially monitored operating unit. Time diagram of functioning process of such a system is presented in Figure 4.15.

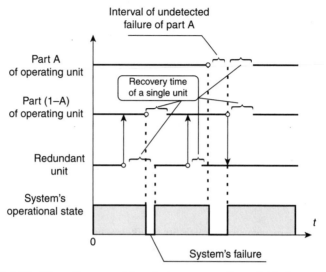

**FIGURE 4.15**  Time diagram of functioning process for a dubbed system with a partially monitored operating unit.

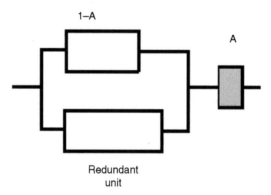

**FIGURE 4.16**  Conditional RBD for a dubbed system with a partially monitored operating unit.

The RBD of such a system can be presented in a conditional form as shown in Figure 4.16.

For such a system, availability coefficient can be easily written as

$$K_{\text{System}} = [1 - (1 - K) \cdot (1 - K_{1-A})] \cdot K_A \approx 1 - \lambda \lambda_{1-A} \tau^2 - \lambda_A \tau).$$

$$(4.22)$$

**TABLE 4.4   Dependence of Recoverable Dubbed System Reliability on the Portion of Uncontrolled Part of the Operating Unit**

| Unit's Availability Coefficient | Portion of Uncontrolled Part | | | |
|---|---|---|---|---|
|  | 0 | 0.01 | 0.05 | 0.1 |
| 0.8 | 0.96 | 0.95 | 0.91 | 0.86 |
| 0.9 | 0.99 | 0.98 | 0.94 | 0.89 |
| 0.95 | 0.9975 | 0.988 | 0.948 | 0.9 |
| 0.99 | 0.9999 | 0.9899 | 0.9499 | 0.8999 |

From (4.22), one can see that even if an uncontrollable part is small enough, the entire effect of redundancy can be practically nullified; moreover, it can even worsen the total system reliability (in Table 4.4 corresponding values are shadowed).

## 4.9   BRIEF HISTORICAL OVERVIEW AND RELATED SOURCES

Here we offer only papers and highly related books to the subject of this chapter. List of general monographs and textbooks, which can include this topic, is given in the main bibliography at the end of the book.

One of the first who suggested applying process of random impulses for reliability analysis of recoverable systems was N. Sedyakin.[7] His ideas were not based on the above-mentioned asymptotic theorems, and therefore were not accepted at that time by mathematical community.

Theory of stochastic point processes and main asymptotic theorems, used for developing the heuristic method, are presented in the following publications. Bibliography is given in chronological–alphabetical order for better exposition of historical background of the subject.

---

[7] Nikolai Mikhailovich Sedyakin (1922–1969) was a Soviet applied mathematician.

## BIBLIOGRAPHY

Cox, D. and W. Smith (1954) On the superposition of renewal processes. *Biometrika*, Vol. 41, Nos. 1–2.

Ososkov, G. A. (1956) A limit theorem for flows of similar events. *Theory Probab. Appl.*, Vol. 1, No. 2.

Renyi, A. (1956) Poisson-folyamat egy jemllemzese. *Ann. Math. Stat.*, Vol. 1, No. 4 (in Hungarian).

Smith, W. L. (1958) Renewal theory and its ramifications. *J. R. Stat. Soc. B*, Vol. 20, pp. 243–302.

Cox, D. R. (1962) *Renewal Theory*. Wiley, New York.

Grigelionis, B. (1963) On convergence of sums of step stochastic processes to a Poisson process. *Theory Probab. Appl.* Vol. 8, No. 2.

Gnedenko, B. V. (1964a) On duplication with renewal. *Eng. Cybern.* No. 5.

Gnedenko, B. V. (1964b) On spare duplication. *Eng. Cybern.* No. 4.

Sedyakin, N. M. (1965) *Elements of the Theory of Random Impulse Flows*. Sovetskoe Radio, Moscow (in Russian).

Cox, D. R. and V. Isham (1980) *Point Processes*. Chapman & Hall.

# 5

# TIME REDUNDANCY

Some systems possess the ability to neglect failures even without redundant units due to some kind of "insensitivity" to short failures or possibility to restart a needed operation. This characteristic can be referred to as "time redundancy."

Let us consider the main types of systems with time redundancy.

## 5.1 SYSTEM WITH POSSIBILITY OF RESTARTING OPERATION

Assume that the needed operation continues for time $t_0$, although it can be completed within some interval $[0, t]$ where $t > t_0$. So, the problem is to find the probability that there is at least one interval between failures that is larger than $t_0$.

There are two cases:

1. Failure durations are negligibly short, although after each failure the system has to begin its operation from the beginning.

*Probabilistic Reliability Models*, First Edition. Igor Ushakov.
© 2012 John Wiley & Sons, Inc. Published 2012 by John Wiley & Sons, Inc.

2. Failure durations are not negligible, so they decrease remaining potentially useful time.

*First Case*   Each failure destroys current results of system operation, and every time the system is forced to begin its operation from the beginning.

Let $R(t|t_0)$ denote the probability that during interval $[0, t]$ there will be at least one period between failures exceeding required value $t_0$. The system performs its operation successfully during time $t$ if two events occur:

- there is no failure in time interval $[0, t_0]$;
- a failure has occurred at $x < t_0$; however, during remaining period $[x, t_0]$, at least once the system successfully performs its operation.

The latter event, evidently, cannot occur if $t - x < t_0$. This verbal explanation leads us to the recurrent expression

$$R(t|t_0) = P(t_0) + \int_0^{t_0} R(t - x|t_0) \mathrm{d}F(x), \qquad (5.1)$$

where $F(x)$, as usual, is d.f of the system TTF.

This is an equation related to Volterra[1]-type equations. Equations of such a recurrent type are usually solved numerically. We will not provide here a mathematical technique for this solution.

For exponentially distributed TTF, one can write a simple approximation based on the following arguments. Successful system operations can occur at the first attempt. If a failure occurs in the first interval, then the system begins the second attempt, and so on. Under condition of a highly reliable system $\lambda t_0 \ll 1$, the probability of appearance of more than one failure within interval of duration $t_0$ is negligibly small. At the same time, the conditional distribution of a single event of the Poisson process has a uniform distribution within a fixed interval.

---

[1] Vito Volterra (1860–1940) was an Italian mathematician and physicist known for his contributions to mathematical biology and integral equations.

Let us explain the latter statement. Consider a Poisson process. If $\lambda t_0 \ll 1$, then the conditional probability that there is a single failure within the considered interval under condition that there is at least one failure is very close to 1. Conditional density of failure location within interval $t$ is

$$f(t | \text{there is exactly one failure}) = \frac{\lambda e^{-\lambda t}}{\lambda t e^{-\lambda t}} = \frac{1}{t}.$$

Thus, for interval $t_0$ on average a failure occurs at moment $t_0/2$, and in this case there remain $t - t_0/2$ units of time for restarting a new attempt. In other words, during time $t$ the system has a possibility to restart operation on average $(t - t_0)/0.5t_0$ times; that is, it is equivalent to corresponding number of loaded redundant units (taking into account that the number of redundant units is integer). Denote the integer part of $(t - t_0)/0.5t_0$ by $\Omega$. Then we can write the following bounds:

$$1 - (q(t_0))^{\Omega} \le P(t | t_0) \le 1 - (q(t_0))^{\Omega+1}. \tag{5.2}$$

If time resource is small, the effect of such time redundancy is negligible. Using formula (5.1) and the same heuristic arguments, we can write for a case when $t < 2t_0$:

$$P(t | t_0) = p(t_0) + q(t - t_0) \cdot p(t_0) = p(t_0) \cdot [1 + q(t - t_0)]. \tag{5.3}$$

Table 5.1 provides some numerical examples where the total operational time is larger in comparison with $t_0$.

*Second Case*  If failures are non-instant, one has to take into account lengths of idle periods between failures. Let $G(t)$ denote a distribution

**TABLE 5.1  Increase of PFFO Depending on Increase of the Total Operational Time**

| $p(t_0)$ | Total Time | | |
| --- | --- | --- | --- |
| | 110% | 120% | 130% |
| 0.99 | 0.991 | 0.992 | 0.993 |
| 0.98 | 0.982 | 0.984 | 0.986 |
| 0.97 | 0.973 | 0.976 | 0.979 |
| 0.96 | 0.964 | 0.968 | 0.972 |
| 0.95 | 0.955 | 0.959 | 0.964 |

of idle time during recovery. This case is very close to the previous one with the difference that restarting of the system occurs after time of recovery, not immediately after a failure.

This verbal description permits us to write the following recurrent expression:

$$R(t|t_0) = P(t_0) + \int_0^{t_0} \left[ \int_0^{t-x} R(t - x - y|t_0) dG(y) \right] dF(x), \qquad (5.4)$$

where again $R(t|t_0) = 0$, if $t < t_0$.

For this case, we cannot suggest any "pleasant result"; solution can be obtained only by numerical methods.

## 5.2  SYSTEMS WITH "ADMISSIBLY SHORT FAILURES"

Consider a system that has some kind of "functional inertia": if recovery time, $\eta$, is less than $\varepsilon$, the system does not "feel" it and continues successfully performing its functions.

It is clear that for highly reliable systems, when $E\{\xi\} \gg E\{\eta\}$, one can apply the asymptotic theorem of point process thinning. If $G(x)$ is d.f. of recovery time, then $G(\varepsilon)$ is the probability that a system failure has been excluded during thinning operation. In other words, such "short" failure has no influence on the system operation. In this case, for general distribution of TTF, $\xi$, we will have a new random variable, $\xi^*$, that has distribution, $F^*(t)$, which is defined as

$$F^*(t) = \begin{cases} F(t) \text{ with probability } G(\varepsilon), \\ F^{*2}(t) = \displaystyle\int_0^t F(t - x) dF(x) \text{ with probability } G(\varepsilon) \cdot [1 - G(\varepsilon)], \\ F^{*3}(t) = \displaystyle\int_0^t F^{*2}(t - x) dF(x) \text{ with probability } G(\varepsilon) \cdot [1 - G(\varepsilon)]^2, \\ F^{*4}(t) = \displaystyle\int_0^t F^{*3}(t - x) dF(x) \text{ with probability } G(\varepsilon) \cdot [1 - G(\varepsilon)]^3, \\ \text{and so on,} \end{cases} \qquad (5.5)$$

where $F^{*n}(t) = \int_0^t F^{*(n-1)}(t - x)dF(x)$ is a convolution of order $n$, that is, distribution of sum of $n$ random variables. Thus, r.v. $\xi^*$ is the sum of random number of random variables $\xi$ and this number has geometric distribution. It is known from the probability theory that asymptotically such sum has an exponential distribution. So, for large $n$ one can approximately write

$$P(t) = \exp\left(-\frac{t}{T} \cdot G(\varepsilon)\right). \tag{5.6}$$

## 5.3    SYSTEMS WITH TIME ACCUMULATION

Some systems accumulate time of successful operation during a total period of performance, $\theta$. The system operation is considered completed if during period $\theta$ the total accumulated operational time exceeds $t_0$. In this case, we consider an alternating process of operating and idle periods.

Denote the probability that the total accumulated operational time is larger than $t_0$ units during period $\theta$ as $P(t_0|\theta)$. For this probability, one considers two events that lead to success:

- a system works without failures during time $t_0$ from the beginning;
- a system has failed at moment $x < t_0$, was repaired during time $y$, and during the remaining interval of $\theta - x - y$ tries to accumulate $t_0 - x$ units of time of a successful operation. This description leads us to the recurrent expression

$$P(t_0|\theta) = P(t_0) + \int_0^{t_0}\left[\int_0^{\theta-x} P(t_0 - x|\theta - x - y)dG(y)\right]dF(x). \tag{5.7}$$

This expression is correct for the case where a system starts to perform at $t = 0$. In general case, such equation can be solved only numerically.

This subject as a whole requires much more room and details. There are many interesting detailed models concerning, for instance, computer systems.

## 5.4 CASE STUDY: GAS PIPELINE WITH AN UNDERGROUND STORAGE

Consider a simplest gas pipeline system with an underground storage, which allows a user to get gas supply during a pipeline breakout.

Let $\eta$ be the pipeline's random idle time with distribution $G(t) = \Pr\{\eta < t\}$ and $\xi$ be its TTF with a distribution $P(t) = \Pr\{\xi < t\}$. The storage volume is $V$. The speed of the storage expenditure (after a pipeline failure) is $\alpha$, and its speed of refilling is $\beta$. A process of expenditure and refilling of the storage is depicted in Figure 5.1. For simplicity, we assume that the storage begins refilling immediately after the pipeline's repair.

The system's failure occurs when a user does not obtain gas (the storage becomes empty). It is clear that due to supply from the storage, a user does not "feel" short failure times of the pipeline.

Assume that pipeline failures occur "not too often" and the probability of the storage's exhaustion during a pipeline's repair is "small enough." (The meaning of the expressions given in quotation marks will be explained below.) Also assume that pipeline MTBF is much larger than the average duration of the pipeline's repair.

Under these assumptions, one can consider the process of the pipeline's disruption occurrences as a renewal stochastic process. An

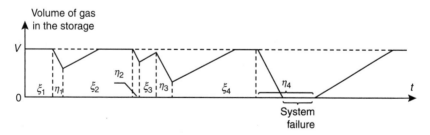

**FIGURE 5.1** Example of a process of expenditure and refilling of the storage.

appearance of the system failures can be considered as a "thinning" procedure because a pipeline's disruption rarely leads to the system's failure. In other words, the process of system failures might be approximated by a Poisson process. Note that for an acceptance of this hypothesis, the probability of developing a pipeline's failure in a system's failure should be small enough (practically less than 0.05).

Let us denote $q = 1 - G(v/\alpha)$, then the system MTTF is

$$T_{\text{syst}} \approx \frac{T + \tau}{1 - G(V/\alpha)}. \tag{5.8}$$

For the PFFO, one can write

$$P_{\text{syst}} \approx \exp\left(-\frac{t}{T_F}\right). \tag{5.9}$$

This expression gives an upper bound because actually we assume that the storage is refilled instantaneously. The smaller the probability $q$, the better the bound.

One can easily obtain a lower bound. Assume that any pipeline failure, which appears during the refilling, leads to the system's failure. This bound is lower because, as a matter of fact, not each failure during refilling leads to the system's failure. The probability of a system's failure under this assumption is

$$q^* = 1 - G\left(\frac{V}{\alpha}\right) + \int_0^x R\left(\frac{\alpha x}{\rho}\right) dG(x). \tag{5.10}$$

Obviously, the probability $q^*$ will be larger if we write

$$q^* = 1 - G\left(\frac{V}{\alpha}\right) + G\left(\frac{V}{\alpha}\right) \cdot R\left(\frac{V}{\rho}\right), \tag{5.11}$$

that is, consider that each pipeline failure has always a maximal duration equal to $V/\rho$. Expression (5.11) can be rewritten as

$$q^* = 1 - G\left(\frac{V}{\alpha}\right)\left[1 - R\left(\frac{V}{\rho}\right)\right]. \tag{5.12}$$

Now we can write a lower bound for the system's MTTF:

$$T^*_{syst} = \frac{T + \tau}{q^*},\tag{5.13}$$

and for the system's PFFO:

$$P^*_{syst} \approx e^{-t/T^*_{syst}}.\tag{5.14}$$

Note that in practical cases values of the MTTF and MTBF coincide.

For more details, see Rudenko and Ushakov (1989).

## 5.5 BRIEF HISTORICAL OVERVIEW AND RELATED SOURCES

Here we offer only papers and highly related books to the subject of this chapter. List of general monographs and textbooks, which can include this topic, is given in the main bibliography at the end of the book.

The so-called time redundancy was first considered in the books by G. Cherkesov and B. Kredentser. Bibliography below is given in chronological–alphabetical order for better exposition of historical background of the subject.

## BIBLIOGRAPHY

Cherkesov, G. N. (1974) *Reliability of Technical Systems with Time Redundancy*. Sovietskoe Radio, Moscow (in Russian).

Kredentser, B. P. (1978) *Prediction of Reliability of Systems with Time Redundancy*. KiNaukova Dumka (in Russian).

Rudenko, Yu. N. and I. A. Ushakov (1989) *Reliability of Energy Systems*. Nauka, Novosibirsk (in Russian).

Obzherin, Yu. E. and A. I. Peschansky (1994) Reliability of unstructured systems with excess time. *Cybern. Syst. Anal.*, Vol. 30, No. 6.

Obzherin, Yu. E. and A. I. Peschansky (2004) Reliability analysis of a system with combined time reserve. *Cybern. Syst. Anal.*, Vol. 40, No. 5.

Obzherin, Yu. E. and A. V. Skatkov (2010) On the time to failure of systems with large replenishable reserve time. *J. Math. Sci.*, Vol. 57, No. 5.

# 6

# "AGING" UNITS AND SYSTEMS OF "AGING" UNITS

## 6.1 CHEBYSHEV BOUND

In practice, very often we know only MTBF of a unit and sometimes, additionally, from some physical hypotheses, that a unit is "aging," that is, its failure rate $\lambda(t)$ is increasing in time. Even this scant information permits to get some reasonable boundary estimates of reliability indices.

If distribution of TTF is unknown, one can write the following upper bound based on the Chebyshev[1] inequality:

$$\Pr\{|\xi - E\{\xi\}| \geq \varepsilon\} \leq \frac{\sigma^2}{\varepsilon^2}, \qquad (6.1)$$

[1] Pafnuty Lvovich Chebyshev (1821–1894) was a Russian mathematician. He is known for his work in the field of probability, statistics, and number theory.

*Probabilistic Reliability Models*, First Edition. Igor Ushakov.
© 2012 John Wiley & Sons, Inc. Published 2012 by John Wiley & Sons, Inc.

where $\xi$ is a random variable (in our case, unit's TTF), and $E\{\xi\}$ and $\sigma^2$ are its mean and variance, respectively; $\varepsilon$ is an arbitrary positive constant. Let us demonstrate the proof of the statement. By definition,

$$\Pr\{|\xi - E\{\xi\}| \geq \varepsilon\} = \int_{|x-E\{\xi\}|\geq\varepsilon} dF(x). \qquad (6.2)$$

Since the domain of integration is $(1/\varepsilon)|x - E\{X\}| \geq 1$, one can write

$$\int_{x-E\{X\}|\geq\varepsilon} dF(x) \leq \frac{1}{\varepsilon^2} \int_{x-E\{X\}|\geq\varepsilon} (-E\{X\})^2 dF(x)$$

$$\leq \frac{1}{\varepsilon^2} \int_{-\infty}^{\infty} (-E\{X\})^2 dF(x) = \frac{\sigma^2}{\varepsilon^2}. \qquad (6.3)$$

This completes the proof.

From (6.3) one can see that the universal estimate gives rather rough estimates and only for arguments that are lying from the mean at a distance larger than $\sigma$. Nevertheless, it is possible to get better estimates if there is some additional information about type of d.f.

## 6.2 "AGING" UNIT

In reliability theory, aging units were introduced by Barlow and Proschan (1965). They called corresponding class of TTF distributions as IFR distributions, where IFR stands for increasing failure rate.

For further deductions, we will use the fact that exponential and degenerate distribution functions represent boundary ones for the entire class of IFR distributions. Indeed, the constant is the boundary function between class of decreasing and increasing functions, and the second one is "the most increasing": it is a delta function.

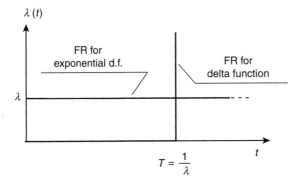

**FIGURE 6.1** Failure rates for exponential and degenerate distributions with equal MTTFs.

Note that degenerate distribution is the distribution function for non-random variable, that is, a constant. Its distribution function is the Heaviside[2] step function defined as

$$D^*(x, t) = \begin{cases} 0, & \text{if } t < x, \\ 1, & \text{otherwise,} \end{cases} \tag{6.4}$$

where $x$ is a point of discontinuity.

In Figure 6.1, failure rates for both cases are depicted.

## 6.3 BOUNDS FOR PROBABILITY OF FAILURE-FREE OPERATIONS

*Lower Bound*   Let us consider three functions: IFR, exponential, and degenerate distributions, all with the same mean equal to $T$. Denote these functions by $P(t)$, $E(t)$, and $D(t)$, respectively (Figure 6.2).

It is easy to show that $t^*$, the crossing point of $P(t)$ and $E(t)$, lies on the right of $T$. From the condition of equality of the means follows that

---

[2] Oliver Heaviside (1850–1925) was a self-taught English electrical engineer, mathematician, and physicist who adapted complex numbers to the study of electrical circuits, invented mathematical techniques to the solution of differential equations (equivalent to Laplace transforms), and independently coformulated vector analysis. Although at odds with the scientific establishment for most of his life, Heaviside changed the face of mathematics and science for years to come.

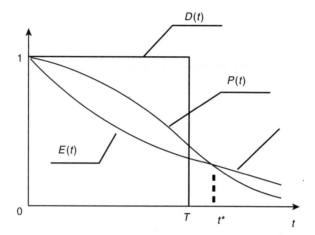

**FIGURE 6.2**    Degenerate, IFR, and exponential distributions with equal MTTFs.

areas restricted by each of the curves $G(t)$, $E(t)$, and $P(t)$ and abscissa are equal. It follows that for $E(t)$ and $P(t)$, the following equality stands:

$$\int_0^{t^*} [P(t) - E(t)]dt = \int_{t^*}^{\infty} [E(t) - P(t)]dt. \tag{6.5}$$

Since $G(t) > P(t)$ for all $t < T$,

$$\int_0^{T} [P(t) - E(t)]dt < \int_0^{T} [G(t) - E(t)]dt = \int_{t^*}^{\infty} [E(t) - P(t)]dt, \tag{6.6}$$

which confirms that inequality $t^* > T$ is satisfied.

From above follows that for any IFR distribution with the mean $T$, the following lower bound exists:

$$P(t) \geq \begin{cases} \exp(-t/T), & \text{for } t < T, \\ 0, & \text{for } t \geq T. \end{cases} \tag{6.7}$$

*Upper Bound*    If $t \leq T$, the upper bound for PFFO is trivial: $P(t) \leq 1$. The upper bound for $t > T$ needs some auxiliary arguments. Consider a family of exponential functions, $E_k^*(t)$, truncated from the right and such that their means are equal to $T$. It is obvious that for each truncated exponent $\lambda_k < \lambda = 1/T$. So, the "aging" function $P(t)$ crosses

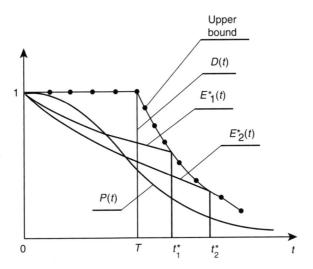

**FIGURE 6.3**    Explanation of the upper bound construction.

any of the truncated exponential functions $E_k^*(t)$ twice: first from
above on the left of $t_k$, and then in $t_k$ (see Figure 6.3). This fact follows
from the condition of the means equality that means that the corresponding areas under curves are equal, that is,

$$\int_{t:P(t)>E_k^*(t)} \left[P(t) - E_k^*(t)\right] dt = \int_{t:E_k^*(t)>P(t)} \left[E_k^*(t) - P(t)\right] dt. \qquad (6.8)$$

For constructing the upper bound, that is set of all $E_k^\&(t_k^*)$. In other
words, for $\int_0^{t_1^*} E_k^\&(t) dt = T$ or each exponential function with parameter $\lambda_k$, one needs to find such $t_k^*$ that

$$\int_0^{t_1^*} E_k^\&(t) dt = \int_0^{t_1^*} \exp(-\lambda_k^\& t) dt = T. \qquad (6.9)$$

Integration of (6.9) gives us

$$1 - \lambda_k^* T = \exp(-\lambda_k^* t_k^*). \qquad (6.10)$$

Thus, we can construct the continuous upper bound for $P(t)$:

$$P(t) \leq \begin{cases} 1, & \text{for } t \leq T, \\ \exp(-\lambda^* t), & \text{for } t > T, \end{cases} \qquad (6.11)$$

where $\lambda^*$ depends on corresponding $t_k$ and is found from equation of type (6.10).

Note that for practical purposes, one is more interested in the lower bound, since it gives a warranty value for PFFO.

## 6.4   SERIES SYSTEM CONSISTING OF "AGING" UNITS

### 6.4.1   Preliminary Lemma

For further analysis, we will need the following lemma.

**Lemma.** Let for functions $f(x)$ and $g(x)$ the following conditions are satisfied:

1. $f(x)$ is a monotone bounded and nonnegative function on positive semi-axis,
2. $g(x)$ is absolutely integrated function on positive semi-axis,
3. $g(x)$ possesses the following property: $g(x) \geq 0$ for $x < a$ and $g(x) \leq 0$ for $x \geq a$, and additionally
4. $g(x)$ satisfies the following condition:

$$\int_0^\infty g(x)dx = 0. \qquad (6.12)$$

Under these conditions, if function $f(x)$ increases (decreases) in $x$, the following inequality takes place:

$$\int_0^\infty f(x)g(x)dx \leq (\geq)0. \qquad (6.13)$$

*Proof.* We present the proof in the form of a chain of equalities:

$$\int_0^\infty f(x)g(x)dx = \int_0^a f(x)g(x)dx + \int_a^\infty f(x)g(x)dx$$

$$\leq (\geq) \int_0^a \left[ \max_{0 \leq x \leq a} f(x) \right] g(x)dx + \int_a^\infty \left[ \min_{0 \leq x \leq a} f(x) \right] g(x)dx$$

$$= f(a) \int_0^a g(x)dx + f(a) \int_a^\infty g(x)dx = f(a) \int_0^\infty g(x)dx = 0.$$

$$(6.14)$$

The common sense of this lemma is easily understood from Figure 6.4: it is clear that in the final integral square $s_2$ is taken with smaller "weight" than square $s_1$.

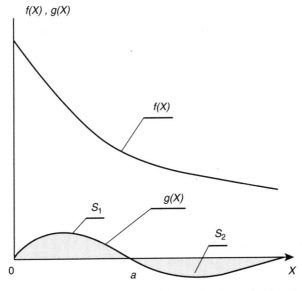

**FIGURE 6.4**   Graphical explanation of lemma for decreasing function $f(x)$.

## 6.5  SERIES SYSTEM

### 6.5.1  Probability of Failure-Free Operation

Consider a series system, elements of which are numerated in order of their increasing MTTFs: $T_1 \leq T_2 \leq \cdots \leq T_n$.

*Lower Bound*    The lower bound for the system, $P(t)$, is defined as the product of the lower bounds of its units:

$$\underline{P}(t) = \begin{cases} E(t) = \displaystyle\prod_{i=1}^{n} E_i(t), & \text{for } t \leq \min_{1\leq i\leq n} T_i, \\ 0, & \text{for } t > \min_{1\leq i\leq n} T_i, \end{cases} \tag{6.15}$$

where $E_i(t) = \exp(-t/T_i)$ and $T_i$ is the $i$th unit MTTF (Figure 6.5).

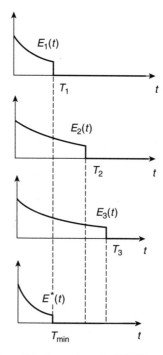

**FIGURE 6.5**  Explanation of the lower bound of PFFO construction for the example of a series system of three units.

Using (6.7), one can immediately write the lower bound for the series system PFFO:

$$\underline{P}(t) \geq \begin{cases} \exp\left(-t\sum_{i=1}^{n}(T_i)^{-1}\right), & \text{for } t < T_{\min}, \\ 0, & \text{for } t \geq T_{\min}. \end{cases} \quad (6.16)$$

Thus, the lower bound for series system PFFO has a nontrivial sense only for time interval $[0, T_{\min}]$.

*Upper Bound*   Using (6.8), one can write the upper bound, $\bar{P}$, for a series system consisting of $n$ independent "aging" units. This time we avoid long formal explanations, referring to Figure 6.6.

The system upper bound for PFFO is defined as

$$\bar{P}^*(t) = \prod_{i=1}^{n} P_i^*(t). \quad (6.17)$$

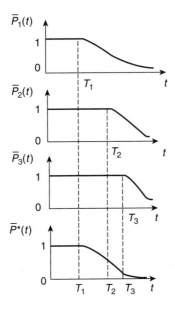

**FIGURE 6.6**   Explanation of the upper bound of PFFO construction for the example of a series system of three units.

After substitution of (6.11) in (6.17), the upper bound expression has the form

$$
\bar{P}(t) = \begin{cases}
1, & \text{for } t \leq T_1, \\
\exp(-\lambda_1 t), & \text{for } T_1 < t \leq T_2, \\
\exp[-(\lambda t_1 + \lambda_2 t)], & \text{for } T_2 < t \leq T_3, \\
\vdots & \\
\exp\left(-\sum_{i=1}^{n} \lambda_i t\right), & \text{for } t \geq T_n,
\end{cases}
\tag{6.18}
$$

where $\lambda_k$ depends on the corresponding moment of truncation.

Unfortunately, this upper bound has almost no practical interest: we are usually interested in PFFO values for $t \ll T_{\min}$.

If each unit's d.f. has a small variation coefficient, then one can assume that

$$
P(t_0) \approx p_{\min}(t_0),
\tag{6.19}
$$

where index "min" corresponds to d.f. with the minimum MTTF, that is,

$$
T_{\min} = \min_{1 \leq k \leq n} T_k.
\tag{6.20}
$$

To explain this in a graphical way, consider a system of two units (Figure 6.7).

From this figure, one can see that in this case one observes "the rule of the weakest link": reliability of the system depends practically only on reliability of the less reliable unit.

### 6.5.2  Mean Time to Failure of a Series System

*Upper Bound*   Since

$$
\underline{P}(t) \leq \min_{1 \leq i \leq n} P_k(t),
\tag{6.21}
$$

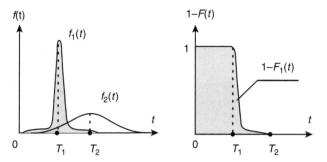

**FIGURE 6.7**  Influence of the weakest unit TTF distribution on PFFO of a series system of two units.

one can write

$$T \le \min_{1 \le i \le n} \int_0^\infty P_i(t)dt = \min_{1 \le i \le n} T_i. \tag{6.22}$$

*Lower Bound*    For getting this bound, use results of the above lemma. Let us show that the substitution of exponential distribution instead of the IFR one (if MTTFs of these units are equal) leads to a decrease in system's MTTF.

Assume that the $n$th "aging" unit is replaced by a unit with exponential distribution of TTF. Calculate the increment of MTTF, $\Delta$:

$$\Delta = \int_0^\infty \prod_{i=1}^n P_i(t)dt - \int_0^\infty \exp\left(-\frac{t}{T_n}\right) \cdot \prod_{i=1}^{n-1} P_i(t)dt$$

$$= \int_0^\infty [P_n(t) - \exp(-t/T_n)] \prod_{i=1}^{n-1} P_i(t)dt. \tag{6.23}$$

Since function $P_n(t)$ crosses function $\exp(-t/T_n)$ only once and from above and their means are equal, the function

$$g(t) = P_n(t) - \exp\left(-\frac{t}{T_n}\right) \tag{6.24}$$

corresponds to function $g(x)$ of the above lemma. At the same time, function $\prod_{i=1}^{n-1} P_i(t)$ corresponds to decreasing function $f(x)$ of the same lemma. Thus, replacing an IFR distribution by an exponential one leads to decrease in the series system MTTF: $\Delta < 0$. Performing such substitutions systematically, one gets the following lower bound for the system MTBF:

$$T_{\text{Syst.}} \geq \int_0^\infty \prod_{i=1}^n \exp\left(-\frac{t}{T_i}\right) dt = \int_0^\infty \exp\left(-t\sum_{i=1}^n \frac{1}{T_i}\right) dt$$

$$= \left(\sum_{i=1}^n \frac{1}{T_i}\right)^{-1}. \tag{6.25}$$

So, the final result can be written as

$$\left(\sum_{i=1}^n \frac{1}{T_i}\right)^{-1} \leq T \leq \min_{1 \leq i \leq n} T_i. \tag{6.26}$$

## 6.6   PARALLEL SYSTEM

### 6.6.1   Probability of Failure-Free Operation

*Upper Bound*   Let us write the formula for the probability of failure of the parallel system:

$$Q(t) = \prod_{i=1}^n q_i(t), \tag{6.27}$$

where $q_i(t)$ is the probability of the $i$th unit failure. Again let us number units in order of their decreasing MTTFs: $T_1 < T_2 < \cdots < T_n$. Note that for the $n$th unit for all $t < T_n$ the upper bound is $\bar{p}_n(t) = 1$, or, equivalently, $\underline{q}_n n(t) = \overline{Q}(t) = 0$. For $t \geq T_n$, all $q_i(t) = 1 - \exp(-\lambda_i^* t)$, where parameters $\lambda_i^*$ were found in (6.10). Thus, for the PFFO of the parallel system consisting of $n$ independent "aging" units, the lower bound, $\underline{Q}(t)$, can be written as

$$\underline{Q}(t) = \begin{cases} 0, & \text{for } t \leq T_n, \\ \prod_{i=1}^n [1 - \exp(-\lambda_i^* t)], & \text{for } t > T_n. \end{cases} \tag{6.28}$$

Naturally, from (6.28), one gets the upper bound for the system PFFO:

$$\bar{P}(t) = \begin{cases} 1, & \text{for } t \le T_n, \\ 1 - \prod_{i=1}^{n}[1 - \exp(-\lambda_i^* t)], & \text{for } t > T_n. \end{cases} \qquad (6.29)$$

Note again, this bound is trivial and noninformative.

If this bound is meaningless, why we pay attention to it? This is given just for information and for protection for those who would like to find such bounds. Remember ancient Greeks who said: "well-competent means armed."

*Lower Bound*   Again, begin with the upper bound for the system failure probability. Use formula (6.27) and pay attention to Figure 6.8 that is, in a sense, a mirror to Figure 6.6.

From this figure, one can see that for $t > T_3 = T_{\max}$ the upper bound is $\bar{Q}(t) = 1$, so the lower bound for PFFO on the same time

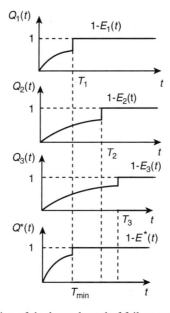

**FIGURE 6.8**   Explanation of the lower bound of failure probability construction for the example of a parallel system of three units.

interval is $\underline{P}(t) = 0$. From Figure 6.8, one can also see that for $t \leq T_1$, the upper bound for the system's failure probability is

$$
\bar{Q}(t) = \begin{cases}
\left[1 - \exp\left(-\dfrac{t}{T_1}\right)\right] \cdot \left[1 - \exp\left(-\dfrac{t}{T_2}\right)\right] \cdot \left[1 - \exp\left(-\dfrac{t}{T_3}\right)\right], & \text{for } 0 < t \leq T_1, \\[2ex]
\left[1 - \exp\left(-\dfrac{t}{T_2}\right)\right] \cdot \left[1 - \exp\left(-\dfrac{t}{T_3}\right)\right], & \text{for } T_1 < t \leq T_2, \\[2ex]
1 - \exp\left(-\dfrac{t}{T_3}\right), & \text{for } T_2 < t \leq T_3, \\[2ex]
1, & \text{for } t > T_3.
\end{cases}
\tag{6.30}
$$

In general case, using the same arguments, one gets

$$
\bar{Q}(t) = \begin{cases}
\displaystyle\prod_{i=1}^{n}[1 - \exp(-t/T_i)], & \text{for } 0 < t \leq T_1, \\[2ex]
\displaystyle\prod_{i=2}^{n}[1 - \exp(-t/T_i)], & \text{for } T_1 < t \leq T_2, \\[1ex]
\;\;\vdots & \\[1ex]
1, & \text{for } t > T_n.
\end{cases}
\tag{6.31}
$$

Naturally, the lower bound for the system PFFO can be obtained as a complement:

$$
\underline{P}(t) = \begin{cases}
1 - \displaystyle\prod_{i=1}^{n}[1 - \exp(-t/T_i)], & \text{for } t \leq T_1, \\[2ex]
1 - \displaystyle\prod_{i=2}^{n}[1 - \exp(-t/T_i)], & \text{for } T_1 < t \leq T_2, \\[1ex]
\;\;\vdots & \\[1ex]
0, & \text{for } t > T_n.
\end{cases}
\tag{6.32}
$$

In conclusion, note that in reliability engineering practice the most important bounds, doubtlessly, are the lower ones, since they give a warranty reliability index value. Fortunately, for "aging" units and systems of "aging" units, these bounds are sufficiently informative.

### 6.6.2  Mean Time to Failure

*Lower Bound*   This bound can be obtained in the usual way:

$$T = \int\limits_0^\infty \left[ 1 - \prod_{i=1}^n q_i(t) \right] dt \geq \max_{1\leq i\leq n} \int\limits_0^\infty [1 - q_i(t)] dt$$

$$= \max_{1\leq i\leq n} \int\limits_0^\infty P_i(t) = T_{\max}. \tag{6.33}$$

*Upper Bound*   Again let us use the result of the above lemma. Let units are numbered in accordance with their increasing MTTFs. The parallel system MTTF is

$$T = \int\limits_0^\infty \left[ 1 - \prod_{i=1}^n q_i(t) \right] dt. \tag{6.34}$$

As above, let us replace the $n$th "aging" unit for a unit with exponential distribution of TTF and the same MTTF:

$$\Delta = \int\limits_0^\infty \left[ 1 - \prod_{i=1}^n q_i(t) \right] dt - \int\limits_0^\infty \left[ 1 - \left[ \left(1 - \exp\left(-\frac{t}{T_n}\right)\right) \cdot \prod_{i=1}^{n-1} q_i(t) \right] \right] dt$$

$$= 1 - \int\limits_0^\infty \prod_{i=1}^{n-1} q_i(t) \cdot \left[ \left(1 - \exp\left(-\frac{t}{T_n}\right)\right) - q_n(t) \right] dt$$

$$= 1 - \int\limits_0^\infty \prod_{i=1}^{n-1} q_i(t) \cdot \left\{ \left[ p_i(t) - \exp\left(-\frac{t}{T_n}\right) \right] \right\} dt. \tag{6.35}$$

The integrand in the last term of (6.35) is a product of two functions. The first one is product of probabilities of unit failures that corresponds to decreasing function $f(x)$ in the above lemma, and the second one is a difference that corresponds to function $g(x)$ in the same lemma. In accordance with the lemma $\Delta < 0$ that means that replacement "aging" unit for a unit with exponential TTF and the same MTTF increases the parallel system MTTF.

Applying such replacement systematically, we finally obtain

$$
T \leq \int_0^\infty \left[ 1 - \prod_{i=1}^H \left( 1 - \exp\left( -\frac{t}{T_i} \right) \right) \right] dt = \int_0^\infty \sum_{i=1}^H \exp\left( -\frac{t}{T_i} \right) dt
$$

$$
- \int_0^\infty \sum_{1 \leq i < j \leq n} \exp\left( -\frac{t}{T_i} \right) \exp\left( -\frac{t}{T_j} \right) dt + \cdots
$$

$$
+ (-1)^{n+1} \int_0^\infty \prod_{i=1}^n \exp\left( -\frac{t}{T_i} \right) dt, \tag{6.36}
$$

which gives, in result, the desired bound.

Thus, MTTF of a parallel system consisting of $n$ independent "aging" units has the following lower and upper bounds:

$$
\max_{1 \leq i \leq m} T_i \leq T_{\text{Syst.}} \leq \sum_{i=1}^m T_i - \sum_{1 \leq i < j \leq m} \left( T_i^{-1} + T_j^{-1} \right)^{-1} + \cdots
$$

$$
+ (-1)^{m+1} \left( \sum_{i=1}^m T_i^{-1} \right)^{-1}. \tag{6.37}
$$

*Remark.* If system's units have IFR distributions of TTF with almost equal MTTFs and very small variation coefficient, then TTF of units could be grouped densely (see Figure 6.9).

In this case, $T_{\min} \approx T_{\max}$ and it means that reliability indices for series and parallel systems are very close! Indeed, in limit case, when all distributions are degenerate and have the same MTTF, we have $\max_{1 \leq i \leq n} T_i = \max_{1 \leq i \leq n} T$ for any $n$. In this case, addition of an extra unit to a series system does not decrease its reliability and addition of an extra unit to a parallel system does not increase its reliability.

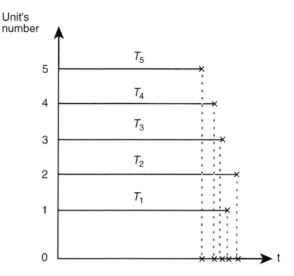

**FIGURE 6.9**   Example of unit TTFs with close values of MTTF and small variation coefficients.


## 6.7   BOUNDS FOR THE COEFFICIENT OF OPERATIONAL AVAILABILITY

We know that the PFFO during interval $t_0$ for a recoverable object that starts its operation at arbitrary random moment $t$ is

$$R(t_0) = \frac{1}{T} \int_t^\infty P(t)\mathrm{d}t, \tag{6.38}$$

where $P(t)$ is unit's PFFO and $T$ is its MTTF.

Use again the above lemma. Consider three functions $P(t)$, $G(t)$, and $E(t)$ introduced above. First take function $g_1(t) = P(t) - G(t)$ that satisfies conditions for function $g(t)$ in the above lemma. Then,

$$\int_{t_0}^\infty [P(t) - G(t)]\mathrm{d}x < 0 \Rightarrow \int_{t_0}^\infty P(t)\mathrm{d}x < \int_{t_0}^\infty G(t)\mathrm{d}x. \tag{6.39}$$

Next take function $g_2(x) = P(t) - G(t)$, for which we have

$$\int_{t_0}^{\infty} [E(t) - P(t)]dx < 0 \Rightarrow \int_{t_0}^{\infty} E(t)dx < \int_{t_0}^{\infty} P(t)dx. \qquad (6.40)$$

From (6.39) and (6.40) follows

$$\int_{t_0}^{\infty} G(t)dx < \int_{t_0}^{\infty} P(t)dx < \int_{t_0}^{\infty} E(t)dx. \qquad (6.41)$$

Since $\int_{t_0}^{\infty} G(t)dt = T - t_0$ and $\int_{t_0}^{\infty} E(t)dt = \int_{t_0}^{\infty} \exp(-t/T)dt$, after substitutions into (6.41), one gets

$$T - t_0 < \int_{t_0}^{\infty} P(t)dt < T \exp(t_0), \qquad (6.42)$$

and, finally,

$$1 - \frac{t_0}{T} < R(t_0) < \exp\left(-\frac{t_0}{T}\right). \qquad (6.43)$$

Obviously, for $t_0 > T$, the left-hand side of inequality turns 0.

Note that bounds (6.43) are extremely good for highly reliable objects. Moreover, for small $t_0$, one can write a simple and very precise approximation

$$R(t_0) \approx 1 - \frac{t_0}{T} \qquad (6.44)$$

Stationary coefficient of interval availability is equal to $R^*(t_0) = KR(t_0)$. It means that for highly reliable objects, one can write

$$R^*(t_0) \approx \left(1 - \frac{t_0}{T}\right)\left(1 - \frac{\tau}{T}\right) \approx 1 - \frac{t_0 + \tau}{T} = 1 - \lambda(t_0 + \tau). \qquad (6.45)$$

Factually, approximation (6.45) is universal for reliability engineering: one should not know the type of unit's TTF distribution. The only condition has to come true: an object (unit or system) has to be "aging," and this condition comes true practically in all real cases.

## 6.8   BRIEF HISTORICAL OVERVIEW AND RELATED SOURCES

Here we offer only papers and highly related books to the subject of this chapter. List of general monographs and textbooks, which can include this topic, is given in the main bibliography at the end of the book.

Swedish engineer, scientist, and mathematician E. Weibull paid attention to the wearing out processes and introduced a two-parameter distribution of rather universal kind. A few years later, B. Gnedenko proved a cycle of limit theorems concerning extreme r.v.'s. The so-called Weibull distribution appeared a particular case of the entire class of limit distributions.

Then in 20 years R. Barlow and F. Proschan developed the modern theory of IFT distributions that appears very constructive in reliability theory.

Bibliography below is given in chronological–alphabetical order for better exposition of historical background of the subject.

## BIBLIOGRAPHY

Weibull, W. (1939) A statistical theory of the strength of materials. *Ing. Vetenskaps Akad. Handl.*, No. 151.

Gnedenko, B. V. (1941) Limit theorems for maximum order statistic. *Rep. Acad. Sci. USSR*, Vol. 32, No. 1 (in Russian)

Gnedenko, B. V. (1943) Sur la Distribution Limite du Terme Maximum d'Une Serie Aleatorie. *Ann. Math.*, Vol. 44, No. 3

Weibull, W. (1951a) A statistical distribution function of wide applicability. *J. Appl. Mech.*, Vol. 18, No. 3.

Weibull, W. (1951b) A statistical distribution function of wide applicability. *J. Appl. Mech.*, Vol. 18, No. 3.

Barlow, R. E., A. W. Marshall, and F. Proschan (1963) Properties of probability distributions with monotone hazard rate. *Ann. Math. Stat.*, Vol. 34.

Barlow, R. E. and A. W. Marshall (1964) Bounds for distributions with monotone hazard rate. I and II. *Ann. Math. Stat.*, Vol. 35.

Barlow, R. E. and A. W. Marshall (1965) Tables of bounds for distributions with monotone hazard rate. *J. Am. Stat. Assoc.*, Vol. 60.

Solovyev, A. D. and I. A. Ushakov (1967) Some bound on a system with aging elements. *Avtom. Vychisl. Tekh.*, Vol. 6 (in Russian)

Raizer, V. (2009) Reliability assessment due to wear. *RTA J.*, Vol. 4, No. 1.

# 7

# TWO-POLE NETWORKS

## 7.1  GENERAL COMMENTS

Earlier we considered systems with the so-called "reducible structure." These are series, parallel, and various kinds of mixtures of series and parallel connections. As mentioned, they are two-pole structures that can be reduced, with the help of a simple routine, to a single equivalent unit. However, not all systems can be described in such a simple way.

We would like to emphasize that most of existing networks, for example, communication and computer networks, transportation railroads, gas and oil pipelines, electric power systems, and others, have a structure that cannot be described in terms of reducible structures.

In general, network reliability can be analyzed from different viewpoints. If a network is designated for transportation of some material flows, then it can be characterized by an ability to deliver required amount of product from sender to receiver. In telecommunication systems, the system has to permit its customers to be able to contact each

*Probabilistic Reliability Models*, First Edition. Igor Ushakov.
© 2012 John Wiley & Sons, Inc. Published 2012 by John Wiley & Sons, Inc.

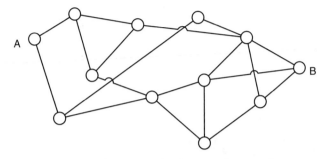

**FIGURE 7.1**   Example of a two-pole network.

other. We will consider two-pole networks where successful operation is characterized by connectivity between points A and B (see Figure 7.1).

It is clear that for two-pole network connectivity necessary and sufficient condition is

- There is at least one *path* from A to B.
  or
- There are no *cuts* between A and B.

Formally, a path is defined as a set of edges that provides connectivity between points A and B. It is also useful to introduce a *minimum path*, that is, such a set of edges that deletion of any of its edges violates network connectivity. A cut is such a set of edges that deletion of all its edges leads to the loss of network connectivity. For cuts, a concept of *minimum cut* is also very important. This is such a set of deleted edges that recovering any of them returns connectivity to the network. Examples of paths and cuts are given in Figure 7.2.

For minimum paths and minimum cuts of a two-pole network, one can introduce the so-called structural functions that, in a sense, repeat concept of series and parallel connections. For minimum path $X^{(\pi)}$, the structural function has the following Boolean form:

$$\phi^{(\pi)}(X^{(\pi)}) = \bigcap_{i \in X^{(\pi)}} x_i, \qquad (7.1)$$

or in verbal explanations:

$$\phi^{(\pi)}(X^{(\pi)}) = \begin{cases} 1, & \text{if and only if all } x_i = 1,\ x_i \in X^{(\pi)}, \\ 0, & \text{otherwise.} \end{cases} \qquad (7.2)$$

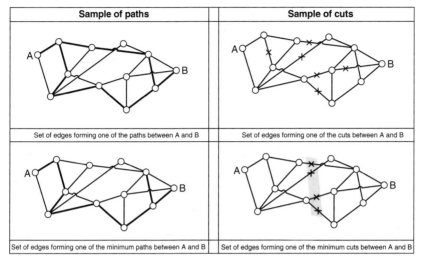

| Sample of paths | Sample of cuts |
|---|---|
| Set of edges forming one of the paths between A and B | Set of edges forming one of the cuts between A and B |
| Set of edges forming one of the minimum paths between A and B | Set of edges forming one of the minimum cuts between A and B |

**FIGURE 7.2**    Examples of paths and cuts of a two-pole network.

For minimum cut $X^{(\kappa)}$, the structural function can be written as

$$\phi^{(\kappa)}(X^{(\kappa)}) = \bigcup_{i \in X^{(\kappa)}} x_i. \qquad (7.3)$$

In verbal form, (7.3) is explained as

$$\phi^{(\kappa)}(X^{(\kappa)}) = \begin{cases} 0, & \text{if and only if all } x_i = 0, \ x_i \in X^{(\kappa)}, \\ 1, & \text{otherwise.} \end{cases} \qquad (7.4)$$

For further convenience, let us introduce symbol $A \Leftrightarrow B$ to denote connectivity between vertices A and B.

### 7.1.1   Method of Direct Enumeration

The simplest example of a system with a nonreducible structure is the so-called "bridge structure" (Figure 7.3).

This particular structure is probably not of great practical importance, but it is reasonable to consider it in order to demonstrate the main methods of analysis of such kinds of structures.

For this simple system, it is possible to enumerate all states and check each of ... (Table 7.1).

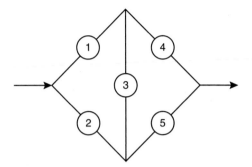

**FIGURE 7.3**    Bridge structure.

**TABLE 7.1    System States Defined by Units' States**

| States of Units | | | | | Vector $X_k$ | Value $\prod(X_k)$ |
|---|---|---|---|---|---|---|
| $x_1$ | $x_2$ | $x_3$ | $x_4$ | $x_5$ | | |
| 1 | 1 | 1 | 1 | 1 | $X_1$ | 1 |
| 0 | 1 | 1 | 1 | 1 | $X_2$ | 1 |
| 1 | 0 | 1 | 1 | 1 | $X_3$ | 1 |
| 1 | 1 | 0 | 1 | 1 | $X_4$ | 1 |
| 1 | 1 | 1 | 0 | 1 | $X_5$ | 1 |
| 1 | 1 | 1 | 1 | 0 | $X_6$ | 1 |
| 0 | 0 | 1 | 1 | 1 | $X_7$ | 0 |
| 0 | 1 | 0 | 1 | 1 | $X_8$ | 1 |
| 0 | 1 | 1 | 0 | 1 | $X_9$ | 1 |
| 0 | 1 | 1 | 1 | 0 | $X_{10}$ | 1 |
| 1 | 0 | 0 | 1 | 1 | $X_{11}$ | 1 |
| 1 | 0 | 1 | 0 | 1 | $X_{12}$ | 1 |
| 1 | 0 | 1 | 1 | 0 | $X_{13}$ | 1 |
| 1 | 1 | 0 | 0 | 1 | $X_{14}$ | 1 |
| 1 | 1 | 0 | 1 | 0 | $X_{15}$ | 1 |
| 1 | 1 | 1 | 0 | 0 | $X_{16}$ | 0 |
| 0 | 0 | 0 | 1 | 1 | $X_{17}$ | 0 |
| 0 | 0 | 1 | 0 | 1 | $X_{18}$ | 0 |
| 0 | 0 | 1 | 1 | 0 | $X_{19}$ | 0 |
| 1 | 0 | 0 | 0 | 1 | $X_{20}$ | 1 |
| 1 | 0 | 0 | 1 | 0 | $X_{21}$ | 0 |
| 1 | 0 | 1 | 0 | 0 | $X_{22}$ | 0 |
| 1 | 0 | 0 | 0 | 1 | $X_{23}$ | 0 |
| 1 | 0 | 0 | 1 | 0 | $X_{24}$ | 1 |
| 1 | 1 | 0 | 0 | 0 | $X_{25}$ | 0 |
| 0 | 0 | 0 | 0 | 1 | $X_{26}$ | 0 |
| 0 | 0 | 0 | 1 | 0 | $X_{27}$ | 0 |
| 0 | 0 | 1 | 0 | 0 | $X_{28}$ | 0 |
| 0 | 1 | 0 | 0 | 0 | $X_{29}$ | 0 |
| 1 | 0 | 0 | 0 | 0 | $X_{30}$ | 0 |
| 0 | 0 | 0 | 0 | 0 | $X_{32}$ | 0 |

If edges of the bridge structure are mutually independent, one can easily find the probability of each state. For instance, probability that state $X_8$ will be realized is $\Pr\{X = X_8\} = \Pr\{x_1 = 0\} \cdot \Pr\{x_2 = 1\} \cdot \Pr\{x_3 = 0\} \cdot \Pr\{x_4 = 1\} \cdot \Pr\{x_5 = 1\} = q_1 p_2 q_1 p_2 p_2$, where $p_k = \Pr\{x_k = 1\}$ and $q_k = 1 - p_k$.

Finally, the probability of bridge connectivity, $P$, can be calculated by the formula

$$P = E\{\phi(X)\} = \left\{ \sum_{1 \le k \le 32} \phi(X_k) \cdot P(X_k) \right\}. \qquad (7.5)$$

Omitting intermediate results, we give the final formula for the connectedness probability (in the case of identical units) in two forms:

$$P = p^5 - 5p^4 + 2p^3 + 2p^2, \qquad (7.6)$$

$$P = 1 - 2q^2 - 2q^3 + 5q^4 - 2q^5. \qquad (7.7)$$

Expression (7.7) is useful in case of highly reliable systems where $q \ll 1$, because it permits to get a simple approximation

$$P \approx 1 - 2q^2. \qquad (7.8)$$

Of course, such a method of direct enumeration allows one to compute the probability of the connectedness of a nonreducible two-pole network only in principle. This method is nonapplicable for problems one meets in practice.

## 7.2  METHOD OF BOOLEAN FUNCTION DECOMPOSITION

Sometimes the method of decomposition of Boolean functions $\prod(X)$ is very effective. Any Boolean function can be decomposed with respect to its argument:

$$\begin{aligned}
\prod(x_1, \ldots, x_k, \ldots, x_n) &= x_k \prod(x_1, \ldots, 1_k, \ldots, x_n) \\
&\cup \bar{x}_k \prod(x_1, \ldots, 0_k, \ldots, x_n),
\end{aligned} \qquad (7.9)$$

or even a set of its arguments (for instance, two of them):

$$\prod(x_1, \ldots, x_k, \ldots, x_n)$$
$$= x_k x_j \prod(x_1, \ldots, 1_k, \ldots, 1_j, \ldots, x_n) \cup \bar{x}_k x_j \prod(x_1, \ldots, 0_k, \ldots, 1_j, \ldots, x_n)$$
$$\cup x_k \bar{x}_j \prod(x_1, \ldots, 1_k, \ldots, 0_j, \ldots, x_n) \cup \bar{x}_k \bar{x}_j \prod(x_1, \ldots, 0_k, \ldots, 0_j, \ldots, x_n),$$
$$(7.10)$$

where we use $1_k$ (or $0_k$) to show that 1 (or 0) is placed in the $k$th position. If we interpret the terms of the Boolean function as events, we can say that these events in (7.9) and (7.10) are mutually exclusive. For instance, the first term in (7.9) includes $x_k$, and the second one includes $\bar{x}_k$. In this case, we can write for (7.9)

$$E\{\prod(x_1, \ldots, x_k, \ldots, x_n)\}$$
$$= E\{x_k \prod(x_1, \ldots, 1_k, \ldots, x_n)\} + E\{\bar{x}_k \prod(x_1, \ldots, 0_k, \ldots, x_n)\}.$$
$$(7.11)$$

Since $x_k$ and $\prod(x_1, \ldots, 1_k, \ldots, x_n)$ are independent as well as and $\Pi(x_1, \ldots, 0_k, \ldots, x_n)$, (7.11) can be finally rewritten in the form

$$E\{\prod(x_1, \ldots, x_k, \ldots, x_n)\}$$
$$= E\{x_k\}E\{\prod(x_1, \ldots, 1_k, \ldots, x_n)\}$$
$$+ E\{\bar{x}_k\}E\{\prod(x_1, \ldots, 0_k, \ldots, x_n)\}.$$
$$(7.12)$$

Let us apply this rule to the bridge structure. Choose unit $x_3$ for decomposition. Then, (7.12) can be rewritten in the concrete form as

$$E\{\prod(x_1, x_2, x_3, x_4, x_5)\}$$
$$= E\{x_3\}E\{\prod(x_1, x_2, 1, x_4, x_5)\} + E\{\bar{x}_3\}E\{\prod(x_1, x_2, 0, x_4, x_5)\}$$
$$= p_3 E\{(x_1 \cup x_2) \cap (x_4 \cup x_5)\} + q_3 E\{(x_1 \cap x_4) \cup (x_2 \cap x_5)\}.$$
$$(7.13)$$

So, we came to a "mixture" of series-parallel and parallel-series systems. It becomes clear with the following explanations. What does it mean that $x_3 = 1$? It means that in the initial bridge structure, unit $x_3$ is absolutely reliable (always in the operational state). Thus, the bridge

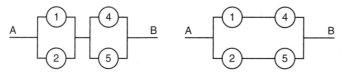

Absolutely reliable unit 3                 Absolutely unreliable unit 3

**FIGURE 7.4** Transformation of initial bridge structure into series-parallel and parallel-series systems depending on state of unit 3.

structure becomes a simple series-parallel structure. Similarly, $x_3 = 0$ means that unit $x_3$ is eliminated from the structure (always in a failed state). This means that the structure becomes a parallel-series structure (see Figure 7.4).

Finally, we may write

$$P = p_3[(1 - q_1q_2)(1 - q_4q_5)] + q_3[1 - (1 - p_1p_4)(1 - p_2p_5)].$$
$$(7.14)$$

We should mention that a Boolean function can be decomposed by any variable. In this particular example, such decomposition can be done with respect to any $x_k$. For example, let us decompose the same structure with respect to unit 1. Corresponding structures are presented in Figure 7.5.

In this case, the expression for the bridge structure connectivity has the form

$$P = p_1\{1 - (1 - q_4) \cdot [1 - (1 - q_3q_2)p_5]\} + q_3p_2[1 - q_5(1 - p_3p_4)].$$
$$(7.15)$$

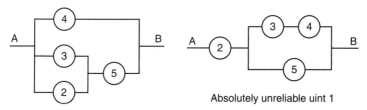

Absolutely reliable uint 1                 Absolutely unreliable uint 1

**FIGURE 7.5** Transformation of initial bridge structure into series-parallel and parallel-series systems depending on state of unit 1.

Of course, (7.14) and (7.15) are equivalent, although (7.14) is much more elegant!

In general, the decomposition method is not practically effective, even if one applies it using decomposition with respect to several Boolean variables. In short, this idea of network decomposition in fact nearly always represents only a nice illustrative example, not an effective tool for engineers.

All of the difficulties connected with the numerical analysis of non-reducible structures lead to a need to find other methods. One effective analytical method is obtaining lower and upper bounds of the unknown value of the probability of connectedness.

## 7.3   METHOD OF PATHS AND CUTS

### 7.3.1   Esary–Proschan Bounds

Consider an arbitrary two-pole network. Assume that all network vertices (or nodes) are absolutely reliable. As we stated above, for two-pole network connectivity at least one minimum path has to exist. Structural function $\prod(X)$ for an arbitrary two-pole network can be written as

$$\phi(X) = \bigcup_{1 \leq k \leq N} \phi_k^{(\pi)}(X). \tag{7.16}$$

Using De Morgan's rule, (7.16) can be rewritten in the form

$$\phi(X) = \bigcup_{1 \leq k \leq N} \phi_k^{(\pi)}(X) = \overline{\bigcap_{1 \leq k \leq N} \overline{\phi_k^{(\pi)}(X)}}, \tag{7.17}$$

where $N$ is the number of all minimum paths of the considered two-pole network.

Factors in (7.17) are mutually dependent, since different paths may have the same links. From the probability theory, we know that if events $X_1$ and $X_2$ are dependent:

$$\Pr\{X_1 \cap X_2\} \neq \Pr\{X_1\}\Pr\{X_2\}. \tag{7.18}$$

In Barlow and Proschan (1965), associated random variables were introduced. Two r.v.'s $X$ and $Y$ are called associated if their covariance is positive:

$$\text{Cov}(X, Y) = E\{X - E\{X\}\} \cdot E\{Y - E\{Y\}\} \geq 0. \qquad (7.19)$$

They naturally generalized this concept on a multivariate case that appeared very useful and productive for reliability analysis of multicomponent complex systems. For associated r.v.'s,

$$\Pr\{X_1 \cap X_2 \cap \cdots \cap X_n\} = \Pr\left\{ \bigcap_{1 \leq k \leq n} X_k \right\} \geq \prod_{1 \leq k \leq n} \Pr\{X_k\}. \qquad (7.20)$$

Note that paths of two-pole network are positively correlated; that is, increased reliability of one of them cannot lead to decreased reliability of another one. Indeed, if reliability of a common unit for both the paths is improved, it improves reliability of both the paths simultaneously. (The same is observed with a decrease in common unit's reliability: there is a decrease in reliability of both the paths.)

So, since for a parallel connection of paths, the probability of disconnection $\prod_{1 \leq k \leq n} \Pr\{X_k = 0\}$ delivers the lower bound, the upper bound for the connectedness of the two-pole network has the form

$$\bar{P} = 1 - \prod_{1 \leq k \leq N} \left(1 - \Pr\{\phi_k^{\pi}(X) = 0\}\right). \qquad (7.21)$$

So, (7.21) is the upper bound for probability of the two-pole network connectivity.

To feel a real sense of all these abstract deductions, let us again demonstrate the method on a bridge structure, presented in Figure 7.3. All possible minimum paths of this structure are given in Figure 7.6.

Expression (7.21) for this particular case takes the form

$$P = 1 - (1 - p^2)^2 \cdot (1 - p^3)^2. \qquad (7.22)$$

We can formulate a similar natural condition of connectedness violation by the following equivalent statement expressed via network's

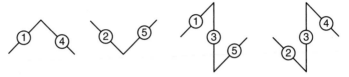

**FIGURE 7.6**   Minimum paths of a bridge structure.

minimum cuts. As we noted above, for violation of the two-pole net-work connectivity links of at least one minimum cut have to be failed. Structural function $\prod(X)$ for an arbitrary two-pole network can be expressed through minimum cuts as follows:

$$\phi(X) = \bigcap_{1 \leq k \leq M} \phi_k^{\kappa}(X), \tag{7.23}$$

where $M$ is the number of all minimum cuts of the considered two-pole network.

Note again that network's minimum cuts may be interdependent because they may contain the same units, and again we observe the positive correlation. Keeping this fact in mind, one can write

$$\underline{P} = \prod_{1 \leq k \leq M} \left(1 - \prod_{i \in \beta_k} q_i\right). \tag{7.24}$$

Formula (7.24) gives the lower bound of value $P$.

Let us illustrate this method on the bridge structure. Minimum cuts of the bridge structure are presented in Figure 7.7.

In Figure 7.8, numerical comparison of the exact values and the upper and lower bounds is given.

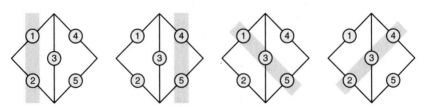

**FIGURE 7.7**   Minimum cuts of a bridge structure.

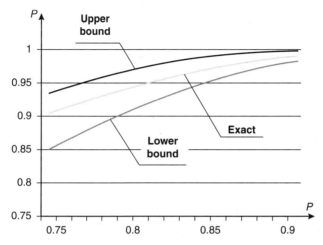

**FIGURE 7.8** Comparison of exact values of bridge structure connectivity and its Esary–Proschan bounds.

### 7.3.2 "Improvements" of Esary–Proschan Bounds

These bounds are based on sets of nonintersecting simple paths and cuts.

We first point out that, for complex networks, the enumeration of all the different simple paths and cuts is a very difficult problem demanding a huge computer memory and an enormous computational time. For systems of any practical dimension, this enumeration problem is essentially impossible.

For this reason, one sometimes attempts to make the computations shorter. On a heuristic level, an explanation of the main idea of such an attempt follows. If we consider a very complex multiunit network, we often can find that a lower bound includes some very "thick" cuts, that is, cuts with a large number of links. It is clear that such a parallel connection is characterized by an extremely high reliability. There is a temptation to exclude such "thick" cuts from consideration: they are very reliable in comparison to the remaining cuts. In other words, the value $1 - q_1 q_2 \cdots q_k$ is so close to 1 that it seems reasonable to replace it with 1. This leads to an increase in the reliability index. Thus, after this, the new lower bound should be even higher than the initial strong lower bound. However, the higher the lower bound, the better it is!

Analogously, the strong upper bound includes some "very long" series connections that may be very unreliable. Again the question arises: why should one take into account such a practically absolutely unreliable series connection of units for computation of a reliability index of a parallel connection? Indeed, for very large $m$, $p_1 p_2 \cdots p_m \to 0$. If one neglects such "very long" paths, the new upper bound becomes lower. This again produces a better upper bound than we have initially!

We must emphasize that such a "heuristic" leads to the very rough mistakes. Indeed, the higher the lower bound, the better it is, but only *if* the lower bound remains a lower bound!

We may obtain strange results using these "simplifications" and "improvements" of the bounds: the obtained "improved" bounds may not be bounds of the unknown value at all! In fact, an "improved" lower bound, obtained in such manner, may be even larger than an "improved" upper bound! Thus, we may obtain new "bounds" that lost all mathematical meaning!

Once more we would like to emphasize that a real heuristic is not an arbitrary guess on an "intuitive level." In our opinion, a heuristic usually must be an "almost proven" simplification of an existing strong solution. Sometimes, instead of a proof one may deliver a set of numerical examples, covering the parametrical area of domain, as a confirmation of the heuristic's validity. Such "experimental mathematics" occupies more and more room in computational methods and very often replaces exact proofs.

Let us illustrate some possible mistakes of using the above-mentioned "simplification" on an example of a bridge structure consisting of identical units. Represent approximation of an upper bound $U = 1 - (1 - p^2)^2 (1 - p^3)^2$ in the form

$$U^* \approx 1 - (1 - p^2)^2, \qquad (7.25)$$

where we keep only the "shortest" minimal paths, and approximation of a lower bound $L = (1 - q^2)^2 (1 - q^3)^2$ in the form

$$L^* \approx (1 - q^2)^2, \qquad (7.26)$$

where we keep only the "thinnest" minimal cuts.

**TABLE 7.2   Comparison of $U^*$ and $L^*$ for Various Values of $p$**

| $p$ | $U^*$ | $L^*$ |
| --- | --- | --- |
| 0.9 | 0.9639 | 0.9801 |
| 0.5 | 0.4375 | 0.5625 |
| 0.1 | 0.0199 | 0.0361 |

It is easy to check that $L^* \geq U^*$. We prefer not to use straight dull transformations to prove this fact. We show this in numerical examples for $p = 0.9$, $p = 0.1$, and $p = 0.5$ (Table 7.2).

We see that "improved" lower bound becomes larger than "improved" upper bound!

### 7.3.3   Litvak–Ushakov Bounds

First of all, remember a very important property of reliability of systems—property of monotonicity. This property is absolutely natural and means, in common terms, that if reliability of any system unit is increased, it cannot lead to a decrease in the system reliability as a whole. Of course, inverse statement is also true: decreasing unit's reliability cannot result in the increase in system reliability. Construction of Litvak–Ushakov bounds is based on this property. The idea is to find a set of independent minimum paths, that is, paths that do not have the same link in different paths.

For explanation of the procedure of constructing such a set, let us begin with an example. In Figure 7.9, the first minimum path (links 1, 4, 8, and 11) is marked with black lines. Remember this path and exclude its entire links (the second figure). In remaining part of two-

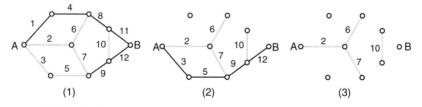

**FIGURE 7.9**   Procedure of constructing nonintersected minimum paths.

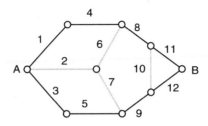

**FIGURE 7.10** Construction of a set of independent minimum paths by exclusion of links 2, 6, 7, and 10.

pole network, find another minimum path (links 3, 5, 9, and 12). After deleting the second path from the initial network, we see that points A and B are disjoint.

This is the end of procedure of finding a set of independent minimum paths: the initial network is reduced to a parallel connection of two minimum paths. Now look at Figure 7.10: one gets the same two paths by excluding the links that are depicted in Figure 7.9(3).

Excluding any link is equivalent to decreasing its reliability to zero. So, the probability of connectivity between nodes A and B for two parallel minimum paths is lower than the same probability for the initial network.

By the way, the number of independent minimum paths cannot be larger than the number of links in the "thinnest" minimum cut. Moreover, if, for instance, the path depicted in Figure 7.11 is chosen, there are no other independent paths in the same two-pole network.

In general case, one can find several sets of minimum paths in the same two-pole network. Each such set forms a parallel connection of

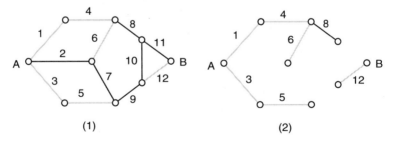

**FIGURE 7.11** An example of a case with a single minimum path.

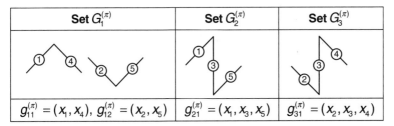

| Set $G_1^{(\pi)}$ | Set $G_2^{(\pi)}$ | Set $G_3^{(\pi)}$ |
|---|---|---|
| $g_{11}^{(\pi)} = (x_1, x_4),\ g_{12}^{(\pi)} = (x_2, x_5)$ | $g_{21}^{(\pi)} = (x_1, x_3, x_5)$ | $g_{31}^{(\pi)} = (x_2, x_3, x_4)$ |

**FIGURE 7.12**   All minimum paths of a bridge structure.

different minimum paths with its own probability of connectivity lower than analogous probability for the initial two-pole network. It is clear that the maximum of lower bounds is the best one. Denote by $G_k^{(\pi)}$ $k$th subset of minimum paths, $k = 1, \ldots, N$, and by $g_{ki}^{(\pi)}$ subset of links consisting of the $i$th path of the $k$th subset. Then the lower Litvak–Ushakov bound is

$$(7.27)$$

Let us again give more detailed explanation of a bridge structure with identical links. All possible minimum paths are depicted in Figure 7.12. The lower bound of Litvak–Ushakov type for the bridge structure is

$$\underline{P} = \max\{[1 - (1 - p^2)^2], p^3, p^3\} = 1 - (1 - p^2)^2. \qquad (7.28)$$

The upper bound can be obtained from a series connection of independent cuts. For explanation of the procedure of constructing the upper bound, let us again refer to the two-pole network presented in Figure 7.9. In Figure 7.13, we demonstrate procedure of sequential construction of a set of independent cuts (each cut here is marked with gray strip). After choosing the first cut, we gather all right ends of

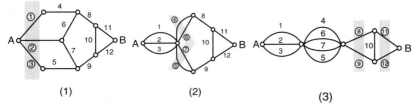

(1)          (2)          (3)

**FIGURE 7.13**   Procedure of constructing nonintersected minimum paths.

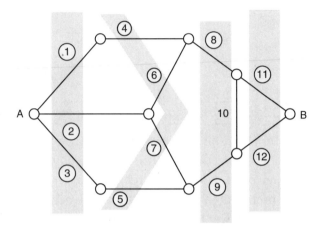

**FIGURE 7.14**   Initial network with marked minimum cuts.

corresponding links into a single node. This procedure is equivalent to putting into the network absolutely reliable links that (by definition of structural monotonicity) can only increase reliability of the entire network.

In the last step in this particular case, we can assume link 10 to be absolutely reliable that completes the construction of this set of independent cuts. The final result can be presented in Figure 7.14.

Note that the number of independent minimum cuts cannot be larger than the number of links in the "longest" minimum path. Moreover, if in Figure 7.15 one chooses cut formed by links 3, 7, 10, and 11, it will be the only cut.

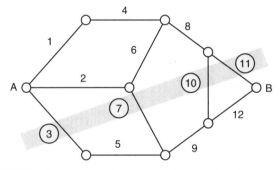

**FIGURE 7.15**   An example of a case with a single minimum path.

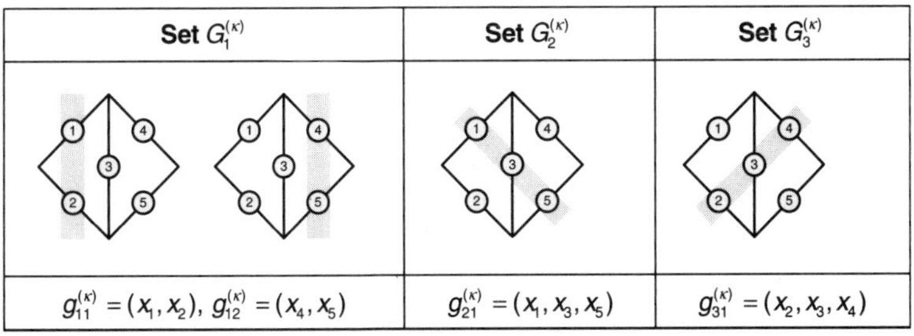

| Set $G_1^{(\varkappa)}$ | Set $G_2^{(\varkappa)}$ | Set $G_3^{(\varkappa)}$ |
|---|---|---|
| | | |
| $g_{11}^{(\varkappa)} = (x_1, x_2),\ g_{12}^{(\varkappa)} = (x_4, x_5)$ | $g_{21}^{(\varkappa)} = (x_1, x_3, x_5)$ | $g_{31}^{(\varkappa)} = (x_2, x_3, x_4)$ |

**FIGURE 7.16**    All minimum cuts of a bridge structure.

In general case, there are several sets of minimum cuts in the same two-pole network. Each such set forms a series connection of different minimum cuts with its own probability of connectivity higher than analogous probability for the initial two-pole network. It is clear that the maximum of lower bounds is the best one. Denote by $G_k^{(\varkappa)}$ $k$th subset of minimum cuts, $k = 1, \ldots, M$, and by $g_{ki}^{(\varkappa)}$ subset of links consisting of the $i$th cut of the $k$th subset. Then the upper Litvak–Ushakov bound can be written as

$$\bar{P} = \min_{1 \leq k \leq M} \left\{ \prod_{i \in G_k^{(\varkappa)}} \left( 1 - \prod_{j \in g_{ki}^{(\varkappa)}} (1 - p_j) \right) \right\}. \qquad (7.29)$$

Let us again give more detailed explanation of a bridge structure with identical links. All possible minimum cuts are depicted in Figure 7.16.

The upper bound of Litvak–Ushakov type for the bridge structure is

$$\bar{P} = \min\{[1 - (1 - p)^2] \cdot [1 - (1 - p)^2], 1 - (1 - p)^3, 1 - (1 - p)^3\}$$
$$= [1 - (1 - p)^2]^2. \qquad (7.30)$$

### 7.3.4    Comparison of the Two Methods

Again we will not misuse mathematical deductions and refer to results of numerical calculations (see Figure 7.17). Here we use subscript EP for Esary–Proschan bounds and subscript LU for Litvak–Ushakov bounds. What we can see from the figure?

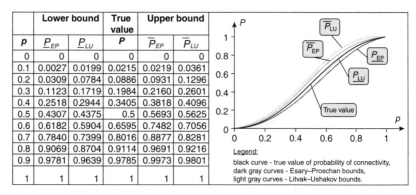

| | Lower bound | | True value | Upper bound | |
|---|---|---|---|---|---|
| **p** | $\underline{P}_{EP}$ | $\underline{P}_{LU}$ | **P** | $\overline{P}_{EP}$ | $\overline{P}_{LU}$ |
| 0 | 0 | 0 | 0 | 0 | 0 |
| 0.1 | 0.0027 | 0.0199 | 0.0215 | 0.0219 | 0.0361 |
| 0.2 | 0.0309 | 0.0784 | 0.0886 | 0.0931 | 0.1296 |
| 0.3 | 0.1123 | 0.1719 | 0.1984 | 0.2160 | 0.2601 |
| 0.4 | 0.2518 | 0.2944 | 0.3405 | 0.3818 | 0.4096 |
| 0.5 | 0.4307 | 0.4375 | 0.5 | 0.5693 | 0.5625 |
| 0.6 | 0.6182 | 0.5904 | 0.6595 | 0.7482 | 0.7056 |
| 0.7 | 0.7840 | 0.7399 | 0.8016 | 0.8877 | 0.8281 |
| 0.8 | 0.9069 | 0.8704 | 0.9114 | 0.9691 | 0.9216 |
| 0.9 | 0.9781 | 0.9639 | 0.9785 | 0.9973 | 0.9801 |
| 1 | 1 | 1 | 1 | 1 | 1 |

Legend:
black curve - true value of probability of connectivity,
dark gray curves - Esary–Proschan bounds,
light gray curves - Litvak–Ushakov bounds.

**FIGURE 7.17**   Comparison of Esary–Proschan and Litvak–Ushakov bounds.

For higher values of unit's reliability index, $p$, the best lower bound is delivered by $\underline{P}_{EP}$. It is important to mention that this bound is one of the most important bounds: for highly reliable two-pole networks, we need to know the warranty level of reliability. (Let us repeat again that if the network is not reliable enough, one should think about reliability improvement, not about reliability evaluation.) Around $p = 0.5$, Esary–Proschan and Litvak–Ushakov bounds cross each other. Then in the area of small values of $p$, the Litvak–Ushakov method delivers the better lower bound. One of the main advantages of the method of obtaining bounds by this method is simplicity of obtaining sets of disjoint paths and cuts.

## 7.4   BRIEF HISTORICAL OVERVIEW AND RELATED SOURCES

Here we offer only papers and highly related books to the subject of this chapter. List of general monographs and textbooks, which can include this topic, is given in the main bibliography at the end of the book.

One of the first works dedicated to the problem of two-pole network connectedness (Moore and Shannon, 1956) was related to connectedness analysis of the so-called "hammock-like" networks. Then series of works by Esary and Proschan (1962, 1963, 1970) were dedicated to obtaining network bounds based on the enumerating of all simple paths and cuts. A decomposition method using this same idea was proposed in Bodin (1970).

The bounds based on disjoint simple paths and cuts were obtained in Ushakov and Litvak 1977 and later were generalized in Litvak and Ushakov 1984. Multiterminal (multipole) networks were investigated in Lomonosov and Polessky 1971.

Bibliography below is given in chronological–alphabetical order for better exposition of historical background of the subject.

## BIBLIOGRAPHY

Moore, E. and C. Shannon (1956) Reliable circuits using less reliable relays. *J. Franklin Inst.*, No. 9.

Esary, J. D. and F. Proschan (1962) The reliability of coherent systems. In: *Redundancy Techniques for Computing Systems*. Spartan Books, Washington, DC.

Ushakov, I. and Yu. Konenkov (1964) Evaluation of branching systems operational effectiveness. In: *Cybernetics for Service to Communism*. Energiya (in Russian).

Kelmans, A. K. (1970) On estimation of probabilistic characteristics of random graphs. *Autom. Remote Control*, Vol. 32, No. 11.

Stepanov, V. E. (1970) On the probability of graph connectivity. *Probab. Theory Appl.*, Vol. 15, No. 1.

Lomonosov, M. V. and V. Polessky (1971) Upper bound for the reliability of information networks. *Problems Inform. Transmission*, Vol. 7.

Lomonosov, M. V. and V. Polessky (1972) Lower bound of network reliability. *Problems Inform. Transmission*, Vol. 8.

Gadasin, V. A. and I. A. Ushakov (1975) *Reliability of Complex Information and Control Systems*. Sovetskoe Radio (in Russian).

Ushakov, I. A. and E. I. Litvak (1977) An upper and lower estimate of the parameters of two-terminal networks. *Eng. Cybern.* Vol. 15.

Litvak, E. and I. Ushakov (1984) Estimation of parameters of structurally complex systems. *Eng, Cybern.*, Vol. 22, No. 4.

Kaustov V. A., E. I. Litvak, and I. A. Ushakov (1986) The computational effectiveness of reliability estimates by the method of nonedge-intersecting chains and cuts. *Sov. J. Comput. Syst. Sci.*, Vol. 24, No. 4.

Reinshke, K. and I. Ushakov (1987) *Application of Graph Theory for Reliability Analysis.* Verlag Technik (in German).

Reinshke, K. and I. Ushakov (1988a) *Application of Graph Theory for Reliability Analysis.* Radio i Sviaz (in Russian).

Reinshke, K. and I. Ushakov (1988b) *Application of Graph Theory for Reliability Analysis.* Springer (in German).

Ushakov, I. (1989) *Reliability Analysis of Computer Systems and Networks.* Mashinostroenie (in Russian).

Elperin, T., I. Gertsbakh, and M. Lomonosov (1991) Estimation of network reliability using graph evolution models. *IEEE Trans. Reliab.*, No. 40.

Lomonosov, M. (1994) On Monte Carlo estimates in network reliability. *Probab. Eng. Inform. Sci.*, No. 8.

Ushakov, I. (1994) *Methods of Research in Telecommunications Reliability (An Overview of Research in the Former Soviet Union).* RTA.

Rauzy, A. (2001) Mathematical foundations of minimal cut sets. *IEEE Trans. Reliab.* Vol. 50, No. 4.

Lin, Y. (2002) Using minimal cuts to evaluate the system reliability of a stochastic-flow network with failures at nodes and arcs. *Reliab. Eng. Syst. Saf.*, Vol. 75.

Fotuhi-Firuzabad, M., R. Billinton, T. S. Munian, and B. Vinayagam (2003) A novel approach to determine minimal tie-sets of complex network. *IEEE Trans. Reliab.*, Vol. 52, No. 4.

Yen, W.-C. (2007) A simple heuristic algorithm for generating all minimal paths. *IEEE Trans. Reliab.*, Vol. 56, No. 3.

# 8

# PERFORMANCE EFFECTIVENESS

## 8.1 EFFECTIVENESS CONCEPTS

Modern large-scale systems are characterized by structural complexity and sophisticated algorithms to facilitate the functioning and interactions of its subsystems. One of the main properties of such systems is that they have a significant "safety factor": even a set of failures may not lead to the system failure, although they can somehow decrease effectiveness of its performance. Indeed, telecommunication networks have highly redundant structures, transportation systems have a number of bypasses, and oil and gas supply systems can change their regime to adjust to new conditions of operation. It allows these systems to perform their operations with lower, though still acceptable, level of effectiveness even with some failed units and subsystems. For such systems, "white-and-black" approach is not appropriate: it is practically impossible to give a definition of system failure. It is more reasonable to say about some degradation of system's abilities; it is natural to speak about performance effectiveness of such systems.

*Probabilistic Reliability Models*, First Edition. Igor Ushakov.
© 2012 John Wiley & Sons, Inc. Published 2012 by John Wiley & Sons, Inc.

In each concrete case, the feature of an index (or indices) of performance effectiveness should be chosen with respect to the type of system
under consideration, its destination, conditions of its operation, and so
on. The physical sense of the performance effectiveness index (PEI) is
usually completely defined by the nature of the system's outcome and
can be evaluated by the same measures. In most practical cases, we can
measure a system's effectiveness in relative units. In other words, we
might take into account the nominal (specified) value of a system's
outcome as the normalizing factor. In other words, a PEI is a measure
of quality and/or volume of its performed functions or operations; that
is, it is a measure of the system's expediency.

Of course, a system's efficiency depends on the type of currently
performed functions and operating environments. A system that is
very efficient under some circumstances might be quite useless and
ineffective under another set of circumstances and/or operations.

If a system's outcome has an upper bound, the PEI can be expressed
in a normalized form; that is, it may be considered as having a positive
value lying between 0 and 1. Then we have $PEI = 0$ if the system has
completely failed and $PEI = 1$ when it is completely operational. For
intermediate states, $0 < PEI < 1$.

Consider a system consisting of $n$ units, each of which can be in two
states: operational and failure. Let $x_i$ be the indicator of the $i$th unit's
state: $x_i = 1$ when the unit is operational and $x_i = 0$ when the unit has
failed. The system has $N = 2^n$ different states determined by the states
of its units. Denote a system state by $X = (x_1, x_2, \ldots, x_n)$.

With time the system changes its state due to changes of states of its
units. Transformation of system states $X(t)$ in time characterizes the
system's behavior. For this state, the performance effectiveness
coefficient is equal to $W_X$, and the system's PEI can be determined as
the expected values of $W_X$:

$$PEI = E\{W_X\} = \sum_{1 \leq k \leq N} W_X P(X). \qquad (8.1)$$

Nevertheless, there are systems for which (8.1) is not valid. For these
systems, effectiveness depends on the entire trajectory of the system's

transition from one state to another during some predetermined time interval. In other words, for these systems the effectiveness is determined by a trajectory of states changing during the system's performance of task.

Examples of such systems are different technological processes, information and computer systems, aircrafts in flight, and so on.

## 8.2  GENERAL IDEA OF EFFECTIVENESS EVALUATION

Let $h_{X_k}(t)$ denote the probability that the system at the moment $t$ is in the state $X_k(t)$. We assume that the current state can be evaluated. Let us denote effectiveness of the system being in state $X$ by $W_X$. It is natural to determine the mathematical expectation of $W_X$ as

$$W_{\text{syst}}(t) = \sum_{1 \le k \le N} h_{X_k}(t) W_{X_k}. \tag{8.2}$$

It is clear that an absolutely accurate calculation of the system's effectiveness when $n \gg 1$ is a difficult, if not unsolvable, computational problem. First of all, one needs to determine a huge number of coefficients $W_k$. Fortunately, it is sometimes not too difficult to split all of the system's states into a relatively small number of classes with close values of $W_k$. If so, we need only to group appropriate states and calculate the corresponding probabilities.

$W_{\text{syst}}$ can then be calculated as

$$W_{\text{syst}}(t) = \sum_{1 \le j \le M} W_j \sum_{X_k \in G_j} h_{X_k}(t), \tag{8.3}$$

where $M$ is the number of different levels of the values of $W_X$ and $G_j$ is the set of system's states, for which $W_X$ belongs to the $j$th level.

Let us evaluate system's effectiveness for a general case. For simplicity of notation, we omit the time $t$ in the expressions below.

Let $h_0$ denote the probability that all units of the system are successfully operating at the moment $t$:

$$h_0 = \prod_{1 \le i \le n} p_i. \tag{8.4}$$

Let $h_i$ denote the probability that only the $i$th unit of the system is in a down state at the moment $t$. Then,

$$h_i = q_i \prod_{1 \le j \le n;\, j \ne i} p_i = \frac{q_i}{p_i} h_0 = g_i h_0, \tag{8.5}$$

where, for brevity, we introduce $g_i = q_i/p_i$, and $h_{ij}$ denotes the probability that only the $i$th and $j$th units of the system are in down states at the moment $t$:

$$h_{ij} = q_i q_j \prod_{1 \le j \le n;\, k \ne (i,j)} p_k = \frac{q_i q_j}{p_i p_j} h_0 = g_i g_j h_0, \tag{8.6}$$

and so on.

We can write the general form of this probability as

$$h_X = \prod_{i \in G_X} p_i \prod_{i \in \bar{G}_X} q_i = h_0 \prod_{i \in \bar{G}_X} g_i, \tag{8.7}$$

where $G_X$ is the set of subscripts of the units, which are considered operational in state $X$, and $\bar{G}_X$ is the complementary set.

Sometimes it is reasonable to write (8.7) for any $X$ as

$$h_X = \prod_{1 \le i \le n_{x_i \in X}} p_i^{x_i} q_i^{(1-x_i)}. \tag{8.8}$$

It is clear that (8.7) and (8.8) are equivalent. Using (8.4)–(8.8), we can rewrite (8.3) as

$$W_{\text{syst}} = W_0 h_0 \left[ 1 + \sum_{1 \le i \le n} \tilde{W}_i g_i + \sum_{1 \le i < j \le n} \tilde{W}_{ij} g_i g_j + \cdots \right], \tag{8.9}$$

where $W_0$ is the system's effectiveness for state $X_0$, and $\tilde{W}_i$, $\tilde{W}_{ij}$, . . .
are normalized effectiveness coefficients for states $X_i$, $X_{ij}$, . . . . In
other words, $\tilde{W}_i = W_i/W_0$, $\tilde{W}_{ij} = W_{ij}/W_0$, and so on.

For a system consisting of highly reliable units, that is, when
$\max_{1 \le i \le n} q_i \ll 1/n$, expression (8.9) can be approximated as

$$W_{\text{syst}} \approx W_0 \left( 1 - \sum_{1 \le i \le n} q_i \right) \cdot \left( 1 + \sum_{1 \le i \le n} \tilde{W}_i q_i \right) \approx W_0 \left( 1 - \sum_{1 \le i \le n} q_i \tilde{w}_i \right).$$

$$(8.10)$$

Here $\tilde{w}_i = 1 - \tilde{W}_i$ has the meaning of a "unit's insignificance."

*Remark.* It is necessary to note that, strictly speaking, it is wrong to
say about "unit's significance." The significance of a unit depends on
the specific state of system. For example, in a simple redundant system
of two units, the significance of any unit is equal to 0 if both units are
successfully operating, but if one unit has failed, then the significance
of the remaining one becomes equal to 1.

### 8.2.1  Conditional Case Study: Airport Traffic Control System

An airport traffic control system consists of two stationary radars, each
with an effective zone of 180° (schematic plot of the system is presented in Figure 8.1). The availability coefficient for each radar is
equal to $K = 0.95$.

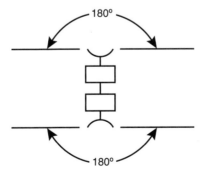

**FIGURE 8.1**   Structure of an airport radar system.

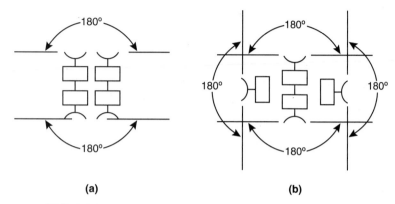

**(a)**                    **(b)**

**FIGURE 8.2**    Two variants of airport radar system improvement.

Assume that if only one radar is operating, the system effectiveness drops to 50% of the nominal value.

It is easy to write the expression for PEI calculation:

$$\text{PEI} = K^2 \cdot 1 + 2K(1 - K) \cdot 0.5 = K^2 + K(1 - K) = K = 0.95. \tag{8.11}$$

There are two variants of the system effectiveness improvement (they are depicted in Figure 8.2). The first variant is a usual redundancy.

In Figure 8.3, arrow shows the direction of radar radiation, and gray color denotes the area covered by radiation of all currently operating radars. With the help of Figure 8.4, one can easily write the expression of PEI for this type of configuration:

$$\text{PEI}^{(1)} = [K^4 + 4K^3(1 - K) + 4K^2(1 - K)^2] \cdot 1 \\ + [2K^2(1 - K)^2 + 4K(1 - K)^3] \cdot 0.5. \tag{8.12}$$

Now consider the second variant where radars form a "ring."

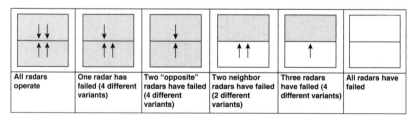

| All radars operate | One radar has failed (4 different variants) | Two "opposite" radars have failed (4 different variants) | Two neighbor radars have failed (2 different variants) | Three radars have failed (4 different variants) | All radars have failed |
|---|---|---|---|---|---|

**FIGURE 8.3**    Possible locations of redundant radar failures.

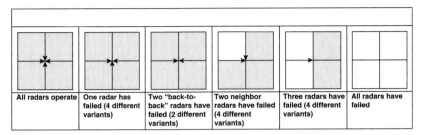

**FIGURE 8.4**  Possible locations of radar failures in case of "ring structure."

Based on Figure 8.4, one can write the following expression:

$$PEI^{(2)} = [K^4 + 4K^3(1 - K) + 2K^2(1 - K)^2] \cdot 1 + 4K^2(1 - K)^2 \cdot 0.75$$
$$+ 4K(1 - K)^3 \cdot 0.5.$$

$$(8.13)$$

These simple analyses show that the second variant is better, although the difference is not too significant:

$$PEI^{(2)} - PEI^{(1)} = K^2(1 - K)^2. \qquad (8.14)$$

For "numerical feeling," in Figure 8.5 comparison of both variants is given.

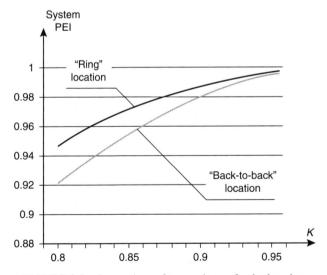

**FIGURE 8.5**  Comparison of two variants of radar location.

## 8.3  ADDITIVE TYPE OF SYSTEM UNITS' OUTCOMES

Let a system consist of $N$ "executive" units, each of which is characterized by its own outcome:

$$w_i = \begin{cases} W_i, & \text{if unit is operable,} \\ 0, & \text{otherwise.} \end{cases} \qquad (8.15)$$

All other units of the system are considered as "administrative" that provide required functioning of executive units. So the system PEI can be represented as the sum of the units' outcomes:

$$\text{PEI} = E\left\{ \sum_{1 \le i \le N} w_i \right\} = \sum_{1 \le i \le N} E\{w_i\}. \qquad (8.16)$$

Expression (8.16) is true even if system's units are dependent. This follows from the well-known fact in mathematical statistics that the expected value of a sum of random variables is equal to the sum of its expected values, regardless of their dependence.

Let a system have the structure shown in Figure 8.6: control center and a set of $N$ executive units monitored from the center. Assume that

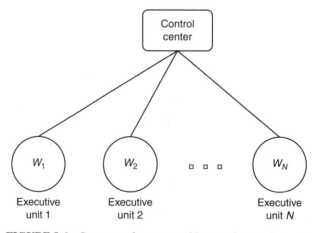

**FIGURE 8.6**   Structure of a system with several executive units.

for successful operation of an executive unit it is necessary that both the center and the unit are operable.

The $i$th executive unit is characterized by its own outcome in operable state, $W_i(X)$, and probability of successful operation, $p_i$, $1 < i < N$. The control center is characterized by its PFFO equal to $p_0$. In this case, the average outcome of the $i$th executive unit is

$$E\{w_i\} = p_0 p_i W_i. \tag{8.17}$$

In this case, a unit's outcome depends on two factors: the operating state of the unit itself and the state of the system. Finally, we can write

$$\text{PEI} = p_0 \sum_{1 \le i \le N} p_i W_i. \tag{8.18}$$

## 8.4   CASE STUDY: ICBM CONTROL SYSTEM

A clear practical example of such a system can be represented by the so-called nonsymmetrical branching system with a simple tree-like hierarchical structure. This system consists of $N$ executive units controlled by "structural" units at higher hierarchy levels. This example is a reminiscent of the author's participation in designing the system for ICBM controlling. Of course, this example presents a schematic and very much conditional structure of the system (see Figure 8.7).

In this figure, headquarter (HQ) controls regional centers (1, 2, ..., 6), and they, in turn, control underground silos with ICBMs. Assume that all silos have PFFO equal to 0.99, headquarter PFFO is 0.995, and regional centers have various PFFOs due to the communication systems, natural environments, and so on. Let the corresponding probabilities for regional centers be $p_1 = 0.995$, $p_2 = 0.993$, $p_3 = 0.992$, $p_4 = 0.99$, $p_5 = 0.89$, and $p_6 = 0.87$. The same system in more formalized form is presented in Figure 8.8.

For such type of a system, a natural measure of importance is the average number of available ICBMs. It is easy to calculate that for given parameters the average of available ICBMs is $\approx 22.45$.

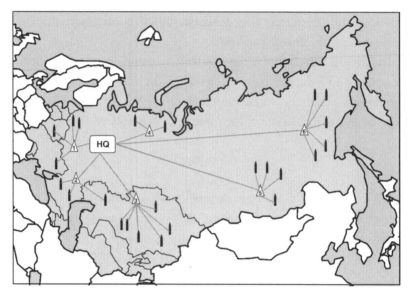

**FIGURE 8.7**    Tree-like hierarchical structure of an ICBM system.

The question could arise: what is the way to increase the system PEI? For instance, what is more effective: to increase the headquarter PFFO from 0.995 to 0.999 or to improve silos and increase the PFFO level up to 0.995?

Simple calculations show that in the first case the system PEI will be ≈22.67, and in the second case it will be ≈22.57. So, the difference in

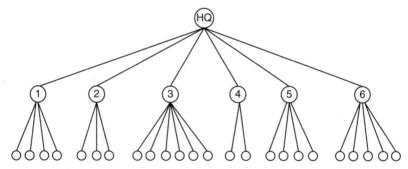

**FIGURE 8.8**    Formalized scheme of the system oriented in Figure 8.7.

two variants is insignificant; however, improving 24 silos, probably, leads to larger expenses.

## 8.5 SYSTEMS WITH INTERSECTING ZONES OF ACTION

Suppose a system consists of $n$ executive units. Unit $i$ has its own zone $Z_i$ of action. Each unit is characterized by its own effectiveness of action $W_i$ in the zone $Z_i$. These zones can be overlapping. Let us denote subzones, $z$, obtained as a result of overlapping by $z$ with subscripts corresponding to the number of zones that form these subzones (see Figure 8.9). Actually, in practice the number of different subzones is restricted enough because overlapping is observed only among neighboring zones.

Such mathematical models appear in the analysis of satellite intelligence systems, in radio communication networks, in power systems, and in anti-aircraft and anti-missile systems (overlapping zones of

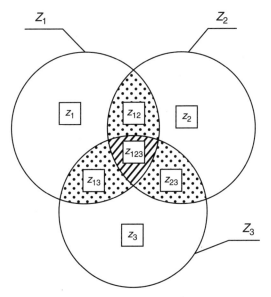

**FIGURE 8.9** Example of a system consisting of three overlapping zones. $Z_1$, $Z_2$, and $Z_3$ are zones and $z_{12}$, $z_{13}$, $z_{23}$, and $z_{123}$ are corresponding subzones.

destruction). Effectiveness of performance within a particular subzone depends on type of the systems and performed operations.

Consider several particular cases.

1. Maximum "intensity index" within subzone.

Assume that each zone individually is characterized by its own "intensity index," $w_k$, per a unit of square. It means that zone $k$ with square $Z_k$ and with no overlapping with other zones has input $w_k Z_k$ in the total system's PEI. If zones $Z_k$ and $Z_j$ overlap, then within the subzone $z_{jk}$ "intensity index" is equal to max$\{w_j, w_k\}$. Subzones with double overlapping are characterized by max$\{w_i, w_j, w_k\}$, and so on.

This mathematical model can be used for describing a reconnaissance satellite (or spy satellite) system. In this case, if a territory is covered by several satellites, the best image is used for further processing.

Calculation algorithm in this case is simple enough, although enumeration remains enumeration. Let us demonstrate the methodology of calculation on a simple example of three zones. Let zone $k$ be operable with probability $p_k$. The system can be in $2^3 = 8$ different states.

First, order all systems in order of value of "intensity indices." Let, for concreteness, $w_1 > w_2 > w_3$. All input data are tabulated in Table 8.1.

**TABLE 8.1    Input Data for Calculation of the System PEI**

|         | $z_1$ | $z_2$ | $z_3$ | $z_{12}$ | $z_{13}$ | $z_{23}$ | $z_{123}$ |
|---------|-------|-------|-------|----------|----------|----------|-----------|
| 1, 1, 1 | $w_1$ | $w_2$ | $w_3$ | $w_1$ | $w_1$ | $w_2$ | $w_1$ |
| 0, 1, 1 | 0     | $w_2$ | $w_3$ | $w_2$ | $w_3$ | $w_2$ | $w_2$ |
| 1, 0, 1 | $w_1$ | 0     | $w_3$ | $w_1$ | $w_1$ | $w_3$ | $w_1$ |
| 1, 1, 0 | $w_1$ | $w_2$ | 0     | $w_1$ | $w_1$ | $w_2$ | $w_1$ |
| 1, 0, 0 | $w_1$ | 0     | 0     | $w_1$ | $w_1$ | 0     | $w_1$ |
| 0, 1, 0 | 0     | $w_2$ | 0     | 0     | 0     | $w_2$ | $w_2$ |
| 0, 0, 1 | 0     | 0     | $w_3$ | 0     | $w_3$ | $w_3$ | $w_3$ |
| 0, 0, 0 | 0     | 0     | 0     | 0     | 0     | 0     | 0     |

The system's performance effectiveness index is calculated by the formula

$$
\begin{aligned}
\text{PEI} = {}& p_1 p_2 p_3 \left[ w_1 (z_1 + z_{12} + z_{13} + z_{123}) + w_2 (z_2 + z_{23}) + w_3 z_3 \right] \\
&+ q_1 p_2 p_3 \left[ w_2 (z_2 + z_{12} + z_{23} + z_{123}) + w_3 (z_3 + z_{13}) \right] \\
&+ p_1 q_2 p_3 \left[ w_1 (z_1 + z_{12} + z_{13} + z_{123}) + w_3 (z_3 + z_{23}) \right] \\
&+ p_1 p_2 q_3 \left[ w_1 (z_1 + z_{12} + z_{13} + z_{123}) + w_2 (z_2 + z_{23}) \right] \\
&+ p_1 q_2 q_3 w_1 (z_1 + z_{12} + z_{13} + z_{123}) \\
&+ q_1 p_2 q_3 w_2 (z_2 + z_{12} + z_{23} + z_{123}) \\
&+ p_1 p_2 q_3 w_3 (z_3 + z_{13} + z_{23} + z_{123}).
\end{aligned}
\tag{8.19}
$$

It is clear that even such a simplified example leads to clumsy calculations. In practice, such systems have more or less homogeneous nature: subzones without overlapping have the same values of $w_i = w^{(0)}$, subzones with a single overlapping also have the same values $w_{ij} = w^{(1)}$, and so on. In addition, zones are of the same size and structure itself is "recurrent," usually honeycomb type (see Figure 8.10).

In this case, it is enough to analyze a single zone $Z_1$ with all its subzones formed with six neighboring zones (Figure 8.11).

However, even in this artificially simplified case, calculation of performance effectiveness index is simple due to enumerating nature of the problem. Let us consider a highly reliable system when probability of occurrence of more than two failures is insignificant. In this case, there are two types of a single failure and four types of two failures (see Figure 8.12).

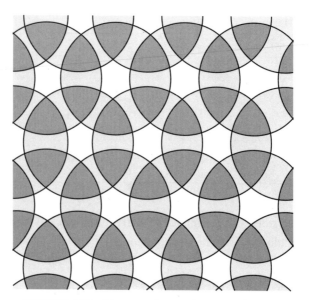

**FIGURE 8.10**  Honeycomb type of zone overlapping.

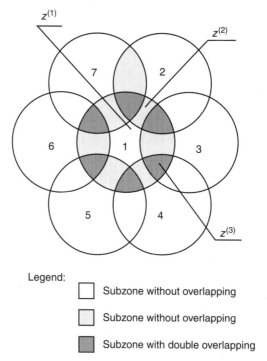

**FIGURE 8.11**  Zone $Z_1$ with its neighboring zones. Here $z^{(k)}$ denotes three types of subzone configurations.

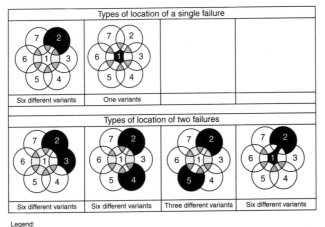

**FIGURE 8.12**   Possible locations of one and two failures.

Based on Figure 8.11, one can write

$$
\begin{aligned}
\text{PEI} \approx\ & p^7\!\left(w^{(1)}z^{(1)} + 6w^{(2)}z^{(2)} + 6w^{(3)}z^{(3)}\right) \\
& + p^6 q\{6[w^{(1)}(z^{(1)} + z^{(2)}) + w^{(2)}(5z^{(2)} + 2z^{(3)}) + 3w^{(3)}z^{(3)}] \\
& + 6w^{(2)}z^{(3)} + 6w^{(1)}z^{(2)})\} \\
& + p^5 q^2\{6[w^{(1)}(z^{(1)} + 2z^{(2)} + z^{(3)}) + w^{(2)}(4z^{(2)} + 2z^{(3)}) + 3w^{(3)}z^{(3)}]\} \\
& + p^5 q^2\{6[w^{(1)}(z^{(1)} + 2z^{(2)}) + w^{(2)}(4z^{(2)} + 4z^{(3)}) + 2w^{(3)}z^{(3)}]\} \\
& + p^5 q^2\{3[w^{(1)}(z^{(1)} + 2z^{(2)}) + w^{(2)}(4z^{(2)} + 4z^{(3)}) + 2w^{(3)}z^{(3)}]\} \\
& + p^5 q^2\{6[w^{(1)}(5z^{(2)} + 2z^{(3)}) + 4w^{(2)}z^{(3)}]\}.
\end{aligned}
\tag{8.20}
$$

Of course, (8.20) can be simplified by collecting terms but we omit primitive though rather boring transformations, keeping in mind that the main target in this case is explanation, not final result.

2. Boolean type of "intensity index" within subzone.

Such a mathematical model can describe an area with a set of ground base stations that serve the cell phone customers: a customer is served or not depending on the zone where he or she is currently in. If a customer is in a zone of intersection of several base stations, one of them is chosen for operation.

This case is similar to the previous one, if we put $w^{(1)} = w^{(2)} = w^{(3)} = w$. Using (8.19), one can immediately write an approximate expression

$$\text{PEI} \approx w\{p^7(z^{(1)} + 6z^{(2)} + 6z^{(3)}) + p^6q[6(z^{(1)}$$
$$+6z^{(2)} + 2z^{(3)}) + 3z^{(3)} + 6z^{(3)}]\}. \qquad (8.21)$$

3. Redundancy type of "intensity index" within subzone.

Imagine that zones represent anti-aircraft or anti-missile areas of defense. In this case, a target in a subzone without overlapping can be shut down with probability $p$, within subzone with overlapping it happens with probability $1 - q^2$, within subzone with double overlapping it happens with probability $1 - q^3$, and so on.

To get approximate expression for PEI, one can use again (8.6), keeping in mind that in this case $w^{(1)} = \rho$, $w^{(2)} = 1 - (1 - \rho)^2$, and $w^{(3)} = 1 - (1 - \rho)^3$.


## 8.6 PRACTICAL RECOMMENDATION

Analysis of the system performance effectiveness must be carried out by a researcher who deeply comprehends the system as a whole, knows its operation, and understands all demands of the system. It is a necessary condition of the successful analysis. Of course, the system analyst should also be acquainted with applying operations research methods. For any operations research problem, the task is concrete and its solution is more of an art than a science.

For the simplicity of discussion, we demonstrate the effectiveness analysis methodology referring to an instant system. The procedure of a system's effectiveness evaluation, roughly speaking, consists of the following tasks:

- formulation of an understandable and clear goal of the system;
- determination of all possible system tasks (operations, functions);
- choice of the most appropriate measure of system effectiveness;
- division of a complex system into subsystems;
- compilation of a structural–functional scheme of the system that reflects the interaction of the system's subsystems;
- collection of reliability data;
- computation of the probabilities of the different states in the system and its subsystems;
- estimation of the effectiveness coefficients of different states;
- performance of the final computations of the system's effectiveness.

We need to remark that the most difficult part of an effectiveness analysis is the evaluation of the coefficients of effectiveness for different system states.

Only in rare cases it is possible to find these coefficients by the means of analytical approaches. The most common method is simulation of the system behavior with the help of computerized models or physical analogues of the system. In the latter case, the analyst introduces different failures at appropriate moments into the system and analyzes the consequences. The last and the most reliable method is to perform experiments with the real system or, at least, with the prototype of the system.

Of course, one has to realize that usually all of these experiments set up to evaluate effectiveness coefficients are very difficult and demand much time, money, and other resources. Consequently, one has to consider how to perform only really necessary experiments. This means that a prior evaluation of different state probabilities is essential: there is no need to analyze extremely rare events.

One can see that an analysis of a system's effectiveness performance is not routine. Designing a mathematical model of a complex system, in some sense, is a problem similar to that of designing the system itself. Of course, there are no technological difficulties, and no time or expense for engineering design and production.

## 8.7    BRIEF HISTORICAL OVERVIEW AND RELATED SOURCES

Here we offer only papers and highly related books to the subject of this chapter. List of general monographs and textbooks, which can include this topic, is given in the general bibliography at the end of the book (Appendix D).

Actually, the main idea of the system's performance effectiveness is provided (with the accuracy of terminology) in Kolmogorov (1945). The first application to reliability theory is described in Ushakov (1960).

Bibliography below is given in chronological–alphabetical order for better exposition of historical background of the subject.

## BIBLIOGRAPHY

Kolmogorov, A. N. (1945) The number of hits in a few shots and general principles for evaluating the effectiveness of the fire. *Proceedings of the Steklov Institute of Mathematics*, Moscow, USSR, Vol. 12, pp. 7–25.

Ushakov, I. A. (1960) Effectiveness estimation of complex systems. In: *Reliability of Electronic Equipment*. Sovetskoe Radio, Moscow.

Ushakov, I. A. and Yu. K. Konenkov (1964) Evaluation of branching systems operational effectiveness. In: *Cybernetics for Service to Communism.* Energiya, Moscow (in Russian).

Ushakov, I. A. (1966) Performance effectiveness of complex systems. In: *On Reliability of Complex Technical Systems.* Sovetskoe Radio, Moscow (in Russian).

Ushakov, I. A. (1967) On reliability of performance of hierarchical branching systems with different executive units. *Eng. Cybern.*, Vol. 5, No. 5.

Ushakov, I. A. and S. Chakravarty (2002a) Reliability influence on communication network capability. *Methods Qual. Manage.*, No. 7.

Ushakov, I. A. and S. Chakravarty (2002b) Reliability measure based on average loss of capacity. *Int. Trans. Oper. Res.*, No. 9.

# 9

# SYSTEM SURVIVABILITY

With the development of extremely complex systems, especially worldwide terrestrial systems, a new problem arose—survivability of systems. Our life is full of unexpected extreme situations: various natural disasters such as earthquakes, floods, forest fires, and so on. The past few decades were marked by evil terrorist actions performed by political terrorists or just maniacs.

Dealing with natural disasters, we need to protect ourselves against the Nature. Here we deal with unpredicted events, and as Albert Einstein[1] said: "God is subtle but he is not malicious."[2] Quite different situation with *Homo sapiens*: he directs his evil will against the most vulnerable and painful points. In a sense, such actions are "more predictable," although it is almost impossible to guess the next step of a crazy maniac or evil suicider.

---

[1] Albert Einstein (1879–1955) was a German-born American theoretical physicist who developed the theory of general relativity.
[2] Inscription in Fine Hall of the Princeton University.

---

*Probabilistic Reliability Models*, First Edition. Igor Ushakov.
© 2012 John Wiley & Sons, Inc. Published 2012 by John Wiley & Sons, Inc.

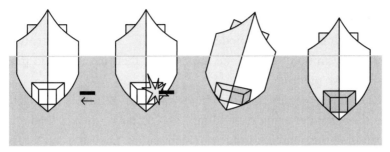

**FIGURE 9.1**    Explanation of Krylov's idea to keep a vessel aswim after enemy torpedo attack.

Taking a historical perspective, we should remember Russian academician Alexei Krylov,[3] who was one of the first who formulated the problem of survivability of sea vessels against enemy attack. He suggested a witty method of keeping a vessel aswim: in response to destruction of some vessel's module, he suggested to fill intentionally a symmetrical module with water and thus keep the vessel in a vertical position (see Figure 9.1). This method of floodability is widely used since then and all military vessels now have module construction of their holds.

The nature of the survivability problem hints that there is no place for pure reliability methods: the destructive events are not predictable in probabilistic sense and there is no such factor as calendar time. Methodology of survivability (or, in inverse terms, vulnerability) is not developed sufficiently by now. However, it is impossible to pass by this problem. Below we will give some simple ideas with no pretensions.

It seems that in this case the most appropriate is minimax approach. In reliability analysis of networks, we already met with the concepts of "minimal cut," that is, with such minimum set of system units that failures lead to the failure of the entire network.

Actually, the only known method of survivability investigation is the so-called "what–if analysis." It is a structured brainstorming method of determining what consequences can happen with the considered

---

[3] Alexei Nikolaevich Krylov (1863–1945) was a Russian and Soviet academician, naval architect, mathematician, and mechanician.

system if one or another situation occurs. Usually, situations under consideration cannot be modeled physically, so the judgments are made by a group of experts on the basis of their intuition and previously known precedents.

One of the natural measurements of survivability is the system effectiveness after a given set of impacts. Of course, this residual effect depends on the number of impacts, their nature, intensity, and location of exposure.

Due to numerical indefinites of expert estimates, comparison of survivability of various systems under some certain impacts is more reasonable than absolute measurements. Consider $N$ various systems designated for performing the same actions (operations). Let each system consist of $n$ units, each of them characterized by its own "level of protection," specified to each system. Denote by $\pi_{ij}$ the level of protection of unit $i$ against impact of type $j$. Assume that there are $n$ possible different impacts of different "destructive levels" $\omega_1, \omega_2, \ldots, \omega_n$. Ascribe to each system unit a loss in case of impact influence, $u_{is}$, where $i = 1, \ldots, n$ and $s = 1, \ldots, N$. These losses can be expressed in absolute (of the same dimension) or relative values. Let $U_s(x_{i_1}, x_{i_2}, \ldots, x_{i_m})$ denote the system loss when set of its units $x_{i_1}, x_{i_2}, \ldots, x_{i_m}$ is destructed. In a sense, the problem is to make such impact assignment that delivers the maximum possible loss for a system. (It will be the worst scenario for "defending side.")

Introduce an additional indicator function of type

$$
\delta_{ip} = \begin{cases} 1, & \text{if } \omega_p > \pi_i, \text{ that is, destructive level larger than protection,} \\ 0, & \text{otherwise.} \end{cases}
$$

For each system, on the basis of input data presented above, one obtains the data provided in Table 9.1.

Thus, in this table some of the values are equal to 0, which means that some impacts are harmless for definite units. Assume that to any unit a single impact can be assigned, that is, $\sum_{1 \le i \le n} \delta_{ip} = 1$ and $\sum_{1 \le p \le n} \delta_{ip} = 1$.

**TABLE 9.1    Loss Value if Impact $j$ Is Applied to Unit $i$ of System $s$**

| Type of Unit | Type of Impact | | | |
|---|---|---|---|---|
| | 1 | 2 | ... | $n$ |
| 1 | $\delta_{11}u_{1s}$ | $\delta_{12}u_{1s}$ | ... | $\delta_{1n}u_{1s}$ |
| 2 | $\delta_{21}u_{2s}$ | $\delta_{22}u_{2s}$ | ... | $\delta_{2n}u_{2s}$ |
| ... | ... | ... | ... | ... |
| $n$ | $\delta_{n1}u_{ns}$ | $\delta_{n2}u_{ns}$ | ... | $\delta_{nn}u_{ns}$ |

Note that there are three general types of systems in sense of their "response" to a set of impacts:

- Systems with linear loss function, that is, such functions that the total loss is equal to the sum of units' losses: $U_s(x_{i_1}, x_{i_2}, \ldots, x_{i_m}) = \sum_{1 \le r \le m} u_{i_r s}$.
- Systems with convex loss function, that is, such functions that the total loss is less than the sum of units' losses: $U_s(x_{i_1}, x_{i_2}, \ldots, x_{i_m}) < \sum_{1 \le r \le m} u_{i_r s}$.
- Systems with concave loss function, that is, such functions that the total loss is larger than the sum of units' losses: $U_s(x_{i_1}, x_{i_2}, \ldots, x_{i_m}) > \sum_{1 \le r \le m} u_{i_r s}$.

Consideration of two last cases in general form has no sense since functions $U_s(x_{i_1}, x_{i_2}, \ldots, x_{i_m})$ should be defined for each specific case. Thus, let us focus on the linear loss function. In this case, the problem of system survivability estimation is reduced to finding

$$U_s = \max_{\delta} \left\{ \sum_{1 \le i \le n} \delta_{ip} u_{is} \,\middle|\, \sum_{1 \le i \le n} \delta_{ip} = 1 \text{ and } \sum_{1 \le p \le n} \delta_{ip} = 1 \right\}.$$

After such calculations for each $s$, $s = 1, \ldots, N$, the final solution is found as $\min_{1 \le s \le N} U_s$.

To make the idea of survivability evaluation more transparent, let us consider a simple numerical example for two simple systems.

**TABLE 9.2   Matrices of Solution for Hostile Impacts on Two Differently Protected Systems**

| The First System | | | | | The Second System | | | | |
|---|---|---|---|---|---|---|---|---|---|
| Level of Protection | Intensity of Impacts | | | | Level of Protection | Intensity of Impacts | | | |
| | 0.5 | 0.65 | 0.8 | 0.95 | | 0.5 | 0.65 | 0.8 | 0.95 |
| 0.3 | 5 | 5 | 5 | 5 | 0.6 | 0 | 7 | 7 | 7 |
| 0.6 | 0 | 7 | 7 | 7 | 0.6 | 0 | 8 | 8 | 8 |
| 0.7 | 0 | 0 | 11 | 11 | 0.7 | 0 | 0 | 11 | 11 |
| 0.9 | 0 | 0 | 0 | 15 | 0.6 | 0 | 12 | 12 | 12 |

## 9.1   ILLUSTRATIVE EXAMPLE

Consider two systems with the following parameters: $u_{11}$, $u_{21}$, $u_{31}$, and $u_{41}$ for the first system and $u_{12}$, $u_{22}$, $u_{32}$, and $u_{42}$ for the second system. Systems have the same "total importance," that is, $u_{11} + u_{21} + u_{31} + u_{41} = u_{12} + u_{22} + u_{32} + u_{42}$, and the "hostile impacts" are of the same intensity. A comparison of survivability of these systems subjected to the same impacts is provided in Table 9.2.

In the table, the cells with chosen units and corresponding impacts are highlighted with gray color. On the basis of this table, one can make some qualitative conclusions. In spite of "more reasonable" location of protection resources in the first system (the more important the unit, the better the protection), the total loss after a hostile attack is 38 conditional units. At the same time, the second system with the same "total importance" and with even allocation of protection resources has the total loss of only 31 conditional units.

Let us consider the second situation: the same total hostile intensity of impacts is distributed more or less evenly. (It can occur, for instance, if terrorists do not know real importance of the system units or if they do not know the level of their protection.) This situation is reflected in Table 9.3.

In this case, the first system is better protected against hostile strike. By the way, disinformation about importance of units and/or levels of protection can help for defending side.

**TABLE 9.3    Matrices of Solution for Hostile Impacts on Two Differently Protected Systems for Even Intensities**

| The First System | | | | | The Second System | | | | |
|---|---|---|---|---|---|---|---|---|---|
| Level of Protection | Intensity of Impacts | | | | Level of Protection | Intensity of Impacts | | | |
| | 0.75 | 0.75 | 0.75 | 0.75 | | 0.75 | 0.75 | 0.75 | 0.75 |
| 0.3 | 5 | 5 | 5 | 5 | 0.6 | 7 | 7 | 7 | 7 |
| 0.6 | 7 | 7 | 7 | 7 | 0.6 | 8 | 8 | 8 | 8 |
| 0.7 | 11 | 11 | 11 | 11 | 0.7 | 11 | 11 | 11 | 11 |
| 0.9 | 0 | 0 | 0 | 0 | 0.6 | 12 | 12 | 12 | 12 |

## 9.2    BRIEF HISTORICAL OVERVIEW AND RELATED SOURCES

One of the first works dedicated to the problem of survivability was Krylov (1942). Almost 40 years later, the problem of survivability arose in connection with vulnerability of large-scale energy systems (Rudenko and Ushakov, 1979, 1989). This problem broke out during the last few years in connection with organized terrorist activity (Ushakov, 2005, 2006; Korczak and Levitin, 2007; Levitin, 2009; Levitin and Ben Haim, 2008, 2009; Levitin and Hausken, 2010).

Bibliography below is given in chronological–alphabetical order for better exposition of historical background of the subject.

## BIBLIOGRAPHY

Krylov, A. N. (1942) *The Ship Theory. Part 1. Ship Stability.* Voenmorizdat.

Rudenko, Yu. N. and I. A.Ushakov (1979) On survivability of complex energy systems. *Energy Syst. Transport.*, No. 1.

Kozlov, M., Yu. Malashenko, V. Rogozhin, I. Ushakov, and T. Ushakova (1986) *Computer Model of Energy Systems Survivability: Methodology, Model, Implementation.* Computer Center of the Russian Academy of Sciences, Moscow.

Rudenko, Yu. N. and I. A. Ushakov (1989) *Reliability of power systems.* Nauka, Novisibirsk (in Russian).

Levitin, G. (2004) Protection survivability importance in systems with multilevel protection. *Qual. Reliab. Eng. Int.*, No. 20.

Ushakov, I. (2005) Cost-effective approach to counter-terrorism. *Commun. Depend. Qual. Manage.*, Vol. 8, No. 3.

Ushakov, I. (2006) Counterterrorism: protection resources allocation. *Reliab. Theory Appl.*, Nos. 2–4.

Bochkov, A. and I. Ushakov (2007) Sensitivity analysis of optimal counter-terrorism resources allocation under subjective expert estimates. *Reliab. Theory Appl.*, Vol. 2, No. 2.

Korczak, E. and G. Levitin (2007) Survivability of systems under multiple factor impact. *Reliab. Eng. Syst. Saf.*, Vol. 92, No. 2.

Levitin, G. and H. Ben Haim (2008) Importance of protections against intentional attacks. *Reliab. Eng. Syst. Saf.*, Vol. 93, No. 4.

Van Gelder, P. (2008) Methods for risk analysis in disaster reduction. *RTA J.*, Vol. 3, No. 2.

Levitin, G. (2009) System survivability and defense against external impacts (Guest editorial). *Int. J. Perform. Eng.*, Vol. 5, No. 1.

Levitin, G. and H. Ben Haim (2009) Minmax defense strategy for complex multi-state systems. *Reliab. Eng. Syst. Saf.*, Vol. 94.

Levitin, G. and K. Hausken (2010) Defense and attack of systems with variable attacker system structure detection probability. *J. Oper. Res. Soc.*, Vol. 61.

# 10

## MULTISTATE SYSTEMS

### 10.1  PRELIMINARY NOTES

Reliability analysis of multistate systems has a long history. The first papers dedicated to this subject appeared as early as 1978 (Barlow and Wu, 1978; El-Neweihi et al., 1978). Later, several papers with introduction of a new technique for multisystem analysis appeared (Ushakov, 1986, 1988, 1998), and finally, a real burst of research papers on the subject (Lisnianski and Levitin, 2003; Levitin et al., 2003; Levitin, 2004, 2005).

We begin the analysis by explaining a new technique introduced in Ushakov (1986).

### 10.2  GENERATING FUNCTION

Instead of constantly attempting to make presentation "transparent" and very much "physical," this time we begin with rather abstract

*Probabilistic Reliability Models*, First Edition. Igor Ushakov.
© 2012 John Wiley & Sons, Inc. Published 2012 by John Wiley & Sons, Inc.

statement of fundamental principles, on which universal generating function (UGF) is based.

Everybody knows about the generating function (GF) that is also called the discrete Laplace transform, or $z$-transform. Generating function is widely used in probability theory, especially for finding convolutions of discrete distributions.

Generating function, $\varphi(z)$, for positive random variable $X$ is defined as the polynomial

$$\varphi(z) = \sum_{1 \le k < \infty} p_k z^{x_k}. \tag{10.1}$$

Power of $z$ denotes the value of r.v., and coefficient at each term is equal to the probability of realization of random variable $x_k$ of the random variable $X$. For instance, for binomial distribution

$$\varphi(z) = \sum_{1 \le k < n} \binom{n}{k} q^k p^{n-k} z^k = (p + qz)^n. \tag{10.2}$$

As one knows, binomial distribution, corresponding to (10.2), characterizes, in particular, a number of failures during testing of $n$ independent and identical items. If one takes two samples of different sizes, say, $n_1$ and $n_2$, from the same general set of events, the GF for such composition will be obtained as the product

$$\varphi_\Sigma(z) = \left[ \sum_{1 \le k < n_1} \binom{n_1}{k} q^k p^{n_1-k} z^k \right] \cdot \left[ \sum_{1 \le j < n_2} \binom{n_2}{k} q^j p^{n_2-j} z^j \right]. \tag{10.3}$$

Let us find the probability that in combined sample with $n_1 = 3$ and $n_2 = 5$ there will have occurred exactly $s = k + j = 3$ failures. For this purpose, compile an auxiliary table of possible outcomes (see Table 10.1).

Theory of various transforms over binomial coefficients is very well developed. In particular, there is known the so-called Vandermonde[1]

---

[1] Alexandre-Théophile Vandermonde (1735–1796) was a French musician, mathematician, and chemist. His name is now principally associated with determinant theory in mathematics.

**TABLE 10.1    Terms Whose Multiplication Leads to $s = 5$**

| Term of the 1$^{st}$ sum in (10.3) | Term of the 2$^{nd}$ sum in (10.3) | Product |
|---|---|---|
| $\binom{3}{0}p^3$ | $\binom{5}{3}q^3p^2$ | $\binom{3}{0}\cdot\binom{5}{3}q^3p^5$ |
| $\binom{3}{1}qp^2$ | $\binom{5}{2}q^2p^3$ | $\binom{3}{1}\cdot\binom{5}{2}q^3p^5$ |
| $\binom{3}{2}q^2p$ | $\binom{5}{1}qp^4$ | $\binom{3}{2}\cdot\binom{5}{1}q^3p^5$ |
| $\binom{3}{3}q^3$ | $\binom{5}{0}p^5$ | $\binom{3}{3}\cdot\binom{5}{0}q^3p^5$ |
| Total probability | | $q^3p^5 \sum_{0\le k\le3}\binom{3}{k}\cdot\binom{5}{3-k}$ |

convolution that in our case has the form

$$\sum_{0\le k\le3}\binom{3}{k}\cdot\binom{5}{5-k}=\binom{8}{3}. \qquad (10.4)$$

So, the term $\binom{8}{3}q^3p^5$ is actually the sixth one in the polynomial expansion

$$\varphi_\Sigma(z)=(p+qz)^3(p+qz)^5=(p+qz)^8$$
$$=p^8+\binom{8}{1}p^7qz+\binom{8}{2}p^6q^2z^2+\binom{8}{3}p^5q^3z^3+\cdots+\binom{8}{7}pq^7z^7+q^8z^8. \qquad (10.5)$$

In a sense, this result was obvious from the beginning: performing a series of two binomial tests of volumes $a$ and $b$ is equivalent to a single test of volume $a+b$.

Of course, similar deductions can be made with other discrete distributions. However, this is not the final target of our discussion.

Working with $z$-transforms "by hand" one uses polynomials because it is convenient to multiply coefficients (probabilities) and add powers at $z$. However, assume that we decided to write a program for computer. What we will do in this case?

**TABLE 10.2    Descartes Product of Two Sets**

|                | $(p_{11}, a_{11})$ | $(p_{12}, a_{12})$ | $\ldots$ | $(p_{15}, a_{15})$ |
|----------------|--------------------|--------------------|----------|--------------------|
| $(p_{21}, a_{21})$ | $(P_{11}, A_{11})$ | $(P_{21}, A_{21})$ | $\ldots$ | $(P_{51}, A_{51})$ |
| $(p_{22}, a_{22})$ | $(P_{12}, A_{12})$ | $(P_{22}, A_{22})$ | $\ldots$ | $(P_{25}, A_{25})$ |
| $\ldots$       | $\ldots$           | $\ldots$           | $\ldots$ | $\ldots$           |
| $(p_{27}, a_{27})$ | $(P_{17}, A_{17})$ | $(P_{27}, A_{27})$ | $\ldots$ | $(P_{27}, A_{27})$ |

We present the first polynomial $\varphi_1(z)$ as a set of pairs $\{(p_{11}, a_{11}),$ $(p_{12}, a_{12}), \ldots, (p_{15}, a_{15})\}$ and the second polynomial $\varphi_2(z)$ as a set of pairs $\{(p_{21}, a_{21}), (p_{22}, a_{22}), \ldots, (p_{25}, a_{25})\}$, where $p_{jk}$'s are corresponding coefficients and $a_{jk}$'s are corresponding powers of polynomials in unfolded form. Then we arrange Descartes product of these two sets (see Table 10.2).

Thus, we have some kind of interaction. Here $P_{jk}$ is found as $P_{jk} = p_{j1} \times p_{2k}$ and $A_{jk}$ is found as $A_{jk} = a_{j1} + a_{2k}$.

However, for a computer there is no difference what kind of operations to perform over the first and the second terms of the considered pairs. This idea has been put on the basis of introducing the so-called universal generating function.

## 10.3    UNIVERSAL GENERATING FUNCTION

We will present UGF only for reliability problems, so we restricted ourselves to units characterized by two parameters: probability of unit's particular state and value of operational parameter associated with this state. Associated parameter can be a value of any system's outcome: productivity rate, capacity, resistance, inductivity, and so on.

In this case, we can keep (just for convenience of using habitual presentation) a polynomial form of specific type: powers of products of two terms, say, $z_a$ and $z_b$, will be presented by some transforms over powers of individual terms, namely,

$$p_a z^a \underset{f}{\otimes} p_b z^b = p_a p_b z^{f(a, b)}, \tag{10.6}$$

where $f$ is an arbitrary given function.

For further discussion, it will be more convenient to use the following form of presentation of (10.6):

$$\{p_a, a\} \underset{f}{\otimes} \{p_b, b\} = \{p_a p_b, f(a, b)\}. \tag{10.7}$$

Naturally, composition operator $\underset{f}{\otimes}$ possesses commutative property, that is,

$$\underset{f}{\otimes}(a, b) = \underset{f}{\otimes}(b, a), \tag{10.8}$$

and associative property, that is,

$$\underset{f}{\otimes}(a, b, c) = \underset{f}{\otimes}(a \underset{f}{\otimes}(b, c)) = \underset{f}{\otimes}((a \underset{f}{\otimes} b), c), \tag{10.9}$$

if the function $f(a, b)$ possesses these properties. In most applications this is the case, although numerous exceptions exist (see, for example, Levitin, 2005).

To avoid terminological confusion, let us call $\underset{f}{\otimes}(a, b, c)$ interaction of variables $a$, $b$, and $c$.

Let us now return back to multistate systems.

Assume that unit $k$ is characterized by the following discrete distribution of its operational parameter $X_k$: $P\{X_k = x_{kj}\} = p_{kj}$. Then we can characterize the distribution of the operational parameter of unit $k$ with the following vector of pairs:

$$\begin{aligned}
\mathbf{Q}_k &= \{(p_{k1}, x_{k1}), (p_{k2}, x_{k2}), \ldots, (p_{ks(k)}, x_{ks(k)})\} \\
&= \{(p_{kj}, x_{kj}), 1 \le j \le s(k)\},
\end{aligned}$$

where $s(k)$ is the number of different values of r.v. $X_k$.

Interaction of operational parameters of two units $X_k$ and $X_i$ can be written as

$$\begin{aligned}
\mathbf{Q}_k \underset{f}{\otimes} \mathbf{Q}_i &= \{(p_{kj}, x_{kj}), 1 \le j \le s(k)\} \underset{f}{\otimes} \{(p_{il}, x_{il}), 1 \le l \le s(i)\} \\
&= (p_{kj} \times p_{il}; f(x_{kj}, x_{il}), j = 1, \ldots, s(k), l = 1, \ldots, s(i)).
\end{aligned}$$

Interaction of operational parameters of $N$ units can be written as

$$\underset{f}{\otimes}(\mathbf{Q}_1, \ldots, \mathbf{Q}_k, \ldots, \mathbf{Q}_N) = \left( \prod_{i=1}^{N} p_{i, m(i)}, f\left(x_{1, m(1)}, \ldots, x_{N, m(N)}\right) \right)$$

$$(10.10)$$

for all combinations of $m(i)$, where $1 \leq m(i) \leq s(i)$.

In expression (10.10), there could be pairs with the same values of operational parameters, for instance, $(P_1, A)$, $(P_2, A)$, $\ldots$, $(P_n, A)$, then the resulting expression (10.10) has to be changed by collecting terms:

$$(P_1, A), (P_2, A), \ldots, (P_n, A) = \left( \sum_{1 \leq k \leq n} P_k, A \right).$$

Let us consider several simple examples for demonstration of the use of UGF.

## 10.4   MULTISTATE SERIES SYSTEM

For demonstration of how UGF works, let us consider several simple numerical examples possessing a transparent physical sense.

### 10.4.1   Series Connection of Piping Runs

Consider a simple oil pipeline (Figure 10.1) consisting of four runs of piping that we will call below as units (just for convenience).

Piping run throughput, $v$, changes randomly due to various external and internal causes. Let pipeline units be characterized by the following distributions of pipe throughputs:

First unit:

$$p_{11} = \Pr\{v = 100\} = 0.8,$$

$$p_{12} = \Pr\{v = 90\} = 0.15,$$

$$p_{13} = \Pr\{v = 80\} = 0.05,$$

$$p_{14} = \Pr\{\text{complete failure}\} \approx 0.$$

**FIGURE 10.1**  Conditional structure of oil pipeline with four series piping runs.

Second unit:

$$p_{21} = \Pr\{v = 100\} = 0.9,$$

$$p_{22} = \Pr\{v = 90\} = 0.07,$$

$$p_{23} = \Pr\{v = 80\} = 0.03,$$

$$p_{24} = \Pr\{\text{complete failure}\} \approx 0.$$

Third unit has the same distribution as the second one.

*Remark*: Pipe run throughout is taken in some conditional units.

The entire pipeline is characterized by minimum current value of its units' throughputs, that is, we have to use the operator $\underset{min}{\otimes}$ because

$$f^{(v)}_{\text{SERIES}}(v_k, v_j) = \min(v_k, v_j).$$

Let us consider the following recurrent procedure.

- *Step 1*. First, consider interaction of parameters of units 1 and 2. Take a Descartes product presenting it in the form of the following table.

|  |  |  | Unit 1 | | | |
|---|---|---|---|---|---|---|
|  |  |  | State 1 (0.8; 100) | State 2 (0.15; 90) | State 3 (0.05; 80) | State 4 (0; 0) |
| Unit 2 | State 1 | (0.9; 100) | 0.8·0.9 = 0.72 min(100, 100) = 100 | 0.15·0.9 = 0.135 min(90, 100) = 90 | 0.05·0.9 = 0.045 min(80, 100) = 80 | (0; 0) |
|  | State 2 | (0.07; 90) | 0.8·0.070 = 0.056 min(100, 90) = 90 | 0.15·0.07 ≈ 0.011 min(90, 90) = 90 | 0.05·0.07 ≈ 0.004 min(80, 90) = 80 | (0; 0) |
|  | State 3 | (0.03; 80) | 0.8·0.03 = 0.024 min(100, 80) = 80 | 0.15·0.03 ≈ 0.005 min(90, 80) = 80 | 0.05·0.03 ≈ 0.002 min(80, 80) = 80 | (0; 0) |
|  | State 4 | (0; 0) | (0; 0) | (0; 0) | (0; 0) | (0; 0) |

In result, one obtains a new equivalent unit with the following distribution of the throughputs:

$$P_1 = \Pr\{v = 100\} = 0.72,$$
$$P_2 = \Pr\{v = 90\} = 0.056 + 0.135 + 0.011 = 0.202,$$
$$P_3 = \Pr\{v = 80\} = 0.024 + 0.005 + 0.002 + 0.004 + 0.045 = 0.082,$$
$$P_4 = \Pr\{v = 0\} = 0.$$

*Remark.* Here and below the sum of all probabilities is not exactly equal to 1 due to rounding of results of multiplications of corresponding probabilities.

- *Step 2.* This new equivalent unit has to be combined with the third unit (see the table below).

| | | | Equivalent Unit | | | |
|---|---|---|---|---|---|---|
| | | | State 1 (0.72; 100) | State 2 (0.202; 90) | State 3 (0.082; 80) | State 4 (0; 0) |
| Unit 3 | State 1 | (0.9; 100) | $0.72 \cdot 0.9 = 0.648$ min(100, 100) = 100 | $0.202 \cdot 0.9 = 0.182$ min(90, 100) = 90 | $0.082 \cdot 0.9 = 0.074$ min(80, 100) = 80 | (0; 0) |
| | State 2 | (0.07; 90) | $0.72 \cdot 0.070 = 0.050$ min(100, 90) = 90 | $0.202 \cdot 0.07 \approx 0.014$ min(90, 90) = 90 | $0.082 \cdot 0.07 \approx 0.006$ min(80, 90) = 80 | (0; 0) |
| | State 3 | (0.03; 80) | $0.72 \cdot 0.03 = 0.022$ min(100, 80) = 80 | $0.202 \cdot 0.03 \approx 0.006$ min(90, 80) = 80 | $0.082 \cdot 0.03 \approx 0.002$ min(80, 90) = 80 | (0; 0) |
| | State 4 | (0; 0) | (0; 0) | (0; 0) | (0; 0) | (0; 0) |

These results allow calculating the expected throughput of the pipeline, $E\{V\}$:

$$E\{V\} = 0.648 \cdot 100 + (0.050 + 0.014 + 0.182) \cdot 90$$
$$+ (0.022 + 0.006 + 0.074 + 0.006 + 0.002) \cdot 80$$
$$= 0.648 \cdot 100 + 0.246 \cdot 90 + 0.011 \cdot 80 \approx 87.8.$$

One can also find the PFFO of this system by some chosen criterion of failure. For instance, if a failure criterion is $V < 90$, then

PFFO is equal to

$$P = \Pr\{V \geq 90\} = 0.648 + 0.246 = 0.894.$$

## 10.4.2  Series Connection of Resistors

Consider a simple chain of ohmic resistors with the structure analogous to that presented in Figure 10.1. Resistance of each unit, $\rho$, can change randomly due to various environmental conditions or internal causes. Let resistors be characterized by the following distributions of resistance:

First unit:

$$p_{11} = \Pr\{\rho = 10\,\text{ohms}\} = 0.8,$$
$$p_{12} = \Pr\{\rho = 9\,\text{ohms}\} = 0.15,$$
$$p_{13} = \Pr\{\rho = 8\,\text{ohms}\} = 0.05,$$
$$p_{14} = \Pr\{\text{complete failure}\} = 0.$$

Second unit:

$$p_{21} = \Pr\{\rho = 10\,\text{ohms}\} = 0.9,$$
$$p_{22} = \Pr\{\rho = 9\,\text{ohms}\} = 0.07,$$
$$p_{23} = \Pr\{\rho = 8\,\text{ohms}\} = 0.3,$$
$$p_{24} = \Pr\{\text{complete failure}\} = 0.$$

Third unit has the same distribution as the second one.

The entire series connection of resistors is characterized by sum of its units' resistances, that is, we have to use the operator $\otimes +$ because in this case

$$f_{\text{SERIES}}^{(\rho)}(\rho_k, \rho_j) = \rho_k + \rho_j.$$

Let us consider the following recurrent procedure.

- *Step 1*. First, consider interaction of parameters of units 1 and 2. Take again a Descartes product presenting it in the form of the following table.

| | | | Unit 1 | | |
|---|---|---|---|---|---|
| | | | State 1 (0.8; 10 ohms) | State 2 (0.15; 9 ohms) | State 3 (0.05; 8 ohms) |
| Unit 2 | State 1 | (0.9; 10 ohms) | 0.8·0.9 = 0.72 <br> 10 + 10 = 20 | 0.15·0.9 = 0.135 <br> 9 + 10 = 19 | 0.05·0.9 = 0.045 <br> 8 + 10 = 18 |
| | State 2 | (0.07; 9 ohms) | 0.8·0.070 = 0.056 <br> 10 + 9 = 19 | 0.15·0.07 ≈ 0.011 <br> 9 + 9 = 18 | 0.05·0.07 ≈ 0.004 <br> 8 + 9 = 17 |
| | State 3 | (0.03; 8 ohms) | 0.8·0.03 = 0.024 <br> 10 + 8 = 18 | 0.15·0.03 ≈ 0.005 <br> 9 + 8 = 17 | 0.05·0.03 ≈ 0.002 <br> 8 + 8 = 16 |

In result, one obtains a new equivalent unit with the following distribution of resistance:

$$P_1 = \Pr\{\rho = 20\,\text{ohms}\} = 0.72,$$
$$P_2 = \Pr\{\rho = 19\,\text{ohms}\} = 0.056 + 0.135 = 0.191,$$
$$P_3 = \Pr\{\rho = 18\,\text{ohms}\} = 0.024 + 0.011 + 0.045 = 0.09,$$
$$P_4 = \Pr\{\rho = 17\,\text{ohms}\} = 0.005 + 0.004 = 0.009,$$
$$P_5 = \Pr\{\rho = 16\,\text{ohms}\} = 0.002.$$

- *Step 2*. This new equivalent unit has to be "converged" with the third unit (see the table below).

| | | | Unit 3 | | |
|---|---|---|---|---|---|
| | | | State 1 (0.9; 10 ohms) | State 2 (0.07; 9 ohms) | State 3 (0.03; 8 ohms) |
| Equivalent unit | State 1 | (0.72; 20 ohms) | 0.9·0.72 = 0.648 <br> 10 + 20 = 30 | 0.07·0.72 ≈ 0.005 <br> 9 + 20 = 29 | 0.03·0.72 = 0.022 <br> 8 + 20 = 28 |
| | State 2 | (0.191; 19 ohms) | 0.9·0.191 = 0.172 <br> 10 + 19 = 29 | 0.07·0.191 = 0.013 <br> 9 + 19 = 28 | 0.03·0.191 = 0.006 <br> 8 + 19 = 27 |
| | State 3 | (0.09; 18 ohms) | 0.9·0.09 = 0.081 <br> 10 + 18 = 28 | 0.07·0.090 ≈ 0.006 <br> 9 + 18 = 27 | 0.03·0.09 = 0.003 <br> 8 + 18 = 26 |
| | State 4 | (0.009; 17 ohms) | 0.9·0.009 ≈ 0.008 <br> 10 + 17 = 27 | 0.07·0.009 ≈ 0 <br> 9 + 17 = 26 | 0.03·0.009 ≈ 0 <br> 8 + 17 = 25 |
| | State 5 | (0.002; 16 ohms) | 0.9·0.002 ≈ 0.002 <br> 10 + 16 = 26 | 0.07·0.002 ≈ 0 <br> 9 + 16 = 25 | 0.03·0.002 ≈ 0 <br> 8 + 16 = 24 |

*Remark.* By the way, this case can be analyzed with a standard GF. We demonstrate it here just for some logical completeness.

These results allow calculating the expected resistance of the series connection of resistors:

$$E\{\rho\} = 0.648 \cdot 30 + (0.172 + 0.005) \cdot 29 + (0.081 + 0.013 + 0.022) \cdot 28$$
$$+ (0.008 + 0.006 + 0.006) \cdot 27 + (0.002 + 0.003) \cdot 26$$
$$= 0.648 \cdot 30 + 0.177 \cdot 29 + 0.116 \cdot 28 + 0.02 \cdot 27 + 0.005 \cdot 26$$
$$\approx 28.49 \text{ ohms.}$$

One can also find the PFFO of this system by some chosen criterion of failure. For instance, if a failure criterion is $\rho < 27$ ohms, then PFFO is equal to

$$P = \Pr\{\rho < 28 \text{ ohms}\} = 0.648 + 0.177 + 0.116 = 0.941.$$

### 10.4.3 Series Connections of Capacitors

Consider a simple chain of ohmic resistors with the structure analogous to that presented in Figure 10.1. Resistance of each unit, $c$, can change randomly due to various environmental conditions or internal causes. Let resistors be characterized by the following distributions of resistance:

First unit:

$$p_{11} = \Pr\{c = 10\ \mu F\} = 0.8,$$

$$p_{12} = \Pr\{c = 9\ \mu F\} = 0.15,$$

$$p_{13} = \Pr\{c = 8\ \mu F\} = 0.05,$$

$$p_{14} = \Pr\{\text{complete failure}\} = 0.$$

Second unit:

$$p_{21} = \Pr\{c = 10\ \mu F\} = 0.9,$$

$$p_{22} = \Pr\{c = 9\ \mu F\} = 0.07,$$

$$p_{23} = \Pr\{c = 8\ \mu F\} = 0.03,$$

$$p_{24} = \Pr\{\text{complete failure}\} = 0.$$

Third unit has the same distribution as the second one.

The entire series connection of capacitors is characterized by sum of inverse values of its units' capacities, that is, we have to use the operator $\overset{\otimes}{f}$, where

$$f^{(c)}_{\text{SERIES}}(c_k, c_j) = (c_k^{-1} + c_j^{-1})^{-1} = \left(\frac{1}{c_k} + \frac{1}{c_j}\right)^{-1} = \frac{c_k \cdot c_j}{c_k + c_j}.$$

Let us consider the following recurrent procedure.

- *Step 1.* First, consider interaction of parameters of units 1 and 2. Take again a Descartes product presenting it in the form of the following table.

| | | | Unit 1 | | |
|---|---|---|---|---|---|
| | | | State 1 | State 2 | State 3 |
| | | | $(0.8; 10\,\mu F)$ | $(0.15; 9\,\mu F)$ | $(0.05; 8\,\mu F)$ |
| Unit 2 | State 1 | $(0.9; 10\,\mu F)$ | $0.8 \cdot 0.9 = 0.72$ | $0.15 \cdot 0.9 = 0.135$ | $0.05 \cdot 0.9 = 0.045$ |
| | | | $(10^{-1} + 10^{-1})^{-1} = 5$ | $(9^{-1} + 10^{-1})^{-1} \approx 4.74$ | $(8^{-1} + 10^{-1})^{-1} \approx 4.44$ |
| | State 2 | $(0.07; 9\,\mu F)$ | $0.8 \cdot 0.070 = 0.056$ | $0.15 \cdot 0.07 \approx 0.011$ | $0.05 \cdot 0.07 \approx 0.004$ |
| | | | $(10^{-1} + 9^{-1})^{-1} \approx 4.74$ | $(9^{-1} + 9^{-1})^{-1} = 4.5$ | $(8^{-1} + 9^{-1})^{-1} \approx 4.24$ |
| | State 3 | $(0.03; 8\,\mu F)$ | $0.8 \cdot 0.03 = 0.024$ | $0.15 \cdot 0.03 \approx 0.005$ | $0.05 \cdot 0.03 \approx 0.002$ |
| | | | $(10^{-1} + 8^{-1})^{-1} \approx 4.44$ | $(9^{-1} + 8^{-1})^{-1} \approx 4.24$ | $(8^{-1} + 8^{-1})^{-1} = 4$ |

In result, one obtains a new equivalent unit with the following distribution of capacity:

$$P_1 = \Pr\{c = 5\,\mu F\} = 0.72,$$
$$P_2 = \Pr\{c = 4.74\,\mu F\} = 0.056 + 0.135 = 0.191,$$
$$P_3 = \Pr\{c = 4.5\,\mu F\} = 0.011,$$
$$P_4 = \Pr\{c = 4.44\,\mu F\} = 0.024 + 0.045 = 0.069,$$
$$P_5 = \Pr\{c = 4.24\,\mu F\} = 0.005 + 0.004 = 0.009,$$
$$P_6 = \Pr\{c = 4\,\mu F\} = 0.002.$$

- *Step 2.* This new equivalent unit has to be "converged" with the third unit (see the table below).

| | | Unit 3 | | |
|---|---|---|---|---|
| | | State 1 | State 2 | State 3 |
| | | $(0.9; 10\,\mu F)$ | $(0.07; 9\,\mu F)$ | $(0.03; 8\,\mu F)$ |
| Equivalent unit | State 1 $(0.72; 5\,\mu F)$ | $0.9 \cdot 0.72 = 0.648$ | $0.07 \cdot 0.72 \approx 0.005$ | $0.03 \cdot 0.72 \approx 0.022$ |
| | | $(10^{-1} + 5^{-1})^{-1} \approx 3.33$ | $(9^{-1} + 5^{-1})^{-1} \approx 3.21$ | $(8^{-1} + 5^{-1})^{-1} \approx 3.08$ |

| State 2 (0.191; | $0.9 \cdot 0.191 = 0.172$ | $0.07 \cdot 0.191 \approx 0.013$ | $0.03 \cdot 0.191 \approx 0.006$ |
|---|---|---|---|
| 4.74 µF) | | | |
| | $(10^{-1} + 4.74^{-1})^{-1} \approx 3.22$ | $(9^{-1} + 4.74^{-1})^{-1} \approx 3.10$ | $(8^{-1} + 4.74^{-1})^{-1} \approx 2.98$ |
| State 3 (0.011; | $0.9 \cdot 0.011 \approx 0.01$ | $0.07 \cdot 0.011 \approx 0.006$ | $0.03 \cdot 0.011 \approx 0.003$ |
| 4.5 µF) | | | |
| | $(10^{-1} + 4.5^{-1})^{-1} \approx 3.10$ | $(9^{-1} + 4.5^{-1})^{-1} = 3$ | $(8^{-1} + 4.5^{-1})^{-1} \approx 2.88$ |
| State 4 (0.069; | $0.9 \cdot 0.069 \approx 0.062$ | $0.07 \cdot 0.069 \approx 0$ | $0.03 \cdot 0.069 \approx 0.002$ |
| 4.44 µF) | | | |
| | $(10^{-1} + 4.44^{-1})^{-1} \approx 3.07$ | $(9^{-1} + 4.44^{-1})^{-1} \approx 2.97$ | $(8^{-1} + 4.44^{-1})^{-1} \approx 2.86$ |
| State 5 (0.009; | $0.9 \cdot 0.009 \approx 0.008$ | $0.07 \cdot 0.009 \approx 0$ | $0.03 \cdot 0.009 \approx 0$ |
| 4.24 µF) | | | |
| | $(10^{-1} + 4.24^{-1})^{-1} \approx 2.98$ | $(9^{-1} + 4.24^{-1})^{-1} \approx 2.88$ | $(8^{-1} + 4.24^{-1})^{-1} \approx 2.77$ |
| State 6 (0.002; | $0.9 \cdot 0.002 \approx 0.002$ | $0.07 \cdot 0.002 \approx 0$ | $0.03 \cdot 0.002 \approx 0$ |
| 4 µF) | | | |
| | $(10^{-1} + 4^{-1})^{-1} \approx 2.86$ | $(9^{-1} + 4^{-1})^{-1} \approx 2.77$ | $(8^{-1} + 4^{-1})^{-1} \approx 2.67$ |

These results allow calculating the expected capacity of the series connection of capacitors:

$$
\begin{aligned}
E\{c\} \approx\ & 0.648 \cdot 3.33 + 0.172 \cdot 3.22 + 0.005 \cdot 3.21 + (0.01 + 0.013) \cdot 3.1 \\
& + 0.022 \cdot 3.08 + 0.062 \cdot 3.07 + 0.006 \cdot 3 + (0.008 + 0.006) \\
& \cdot 2.98 + 0.003 \cdot 2.88 + (0.002 + 0.002) \cdot 2.86 \\
\approx\ & 3.14.
\end{aligned}
$$

One can also find the PFFO of this system by some chosen criterion of failure. For instance, if a failure criterion is $c < 3$ µF, then PFFO is equal to

$$
\begin{aligned}
P = \Pr\{c < 3\ \mu F\} =\ & 0.648 + 0.172 + 0.005 + 0.023 + 0.022 \\
& + 0.062 + 0.006 = 0.938.
\end{aligned}
$$

## 10.5    MULTISTATE PARALLEL SYSTEM

### 10.5.1    Parallel Connection of Piping Runs

Consider a section of oil pipeline with four parallel piping runs (Figure 10.2).

The entire pipeline is characterized by sum of current values of its units' throughputs, $v$, that is, we have to use the operator $\overset{\otimes}{+}$ because

$$
f_{\text{PARALLEL}}^{(p)}(v_1, v_2) = v_1 + v_2.
$$

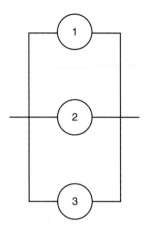

**FIGURE 10.2**    Conditional structure of oil pipeline with three parallel piping runs.

It means that from methodological point of view calculations coincide with those in Section 10.4.2.

### 10.5.2   Parallel Connection of Resistors

Consider a parallel connection of ohmic resistors (see Figure 10.2). For a parallel connection of two resistors, one should use the following function that determines the resistance of the pair of resistors:

$$f^{(\rho)}_{\text{PARALLEL}}(\rho_1, \rho_2) = \frac{\rho_1 + \rho_2}{\rho_1 \cdot \rho_2} \qquad (10.11)$$

and the corresponding operator $\overset{\otimes}{\underset{f}{}}$.

Formally, this mathematical model coincides with that described in Section 10.4.3. One can use all numerical solutions from there with corresponding change of dimension.

### 10.5.3   Parallel Connections of Capacitors

Consider a parallel connection of electrical capacitors (see Figure 10.2). For a parallel connection of two capacitors, one can write

$$f^{(c)}_{\text{PARALLEL}}(c_1, c_2) = c_1 + c_2 \qquad (10.12)$$

and use the operator $\overset{\otimes}{\underset{+}{}}$.

Formally, this type of maniple interaction corresponds to that in Section 10.4.1, so one can use all numerical solutions from that section with corresponding change of dimension.

## 10.6 REDUCIBLE SYSTEMS

Above only simple series and parallel structures were considered. Naturally, the UGF method can be with the same success applied to reducible systems in general. Since routine transforms in this case are very much similar to those described above, we will present only principal new ideas demonstrating them on a simple particular example.

Let a considered system has the structure presented in Figure 10.3.

The UGF for such structure is compiled in accordance with the step-by-step reduction of the initial structure by aggregating pairs of its modules (elements) into single equivalent modules. In other words, the structure is reduced sequentially as presented in Figure 10.4.

Note that in this case one has to use two types of $\underset{f}{\otimes}$ operators with functions $f$ corresponding to series and to parallel connection of structural modules. We will illustrate the methodology on two simple illustrative examples of a pipeline.

**Example 10.1**
Consider a pipeline with a simple reducible structure presented in Figure 10.5.

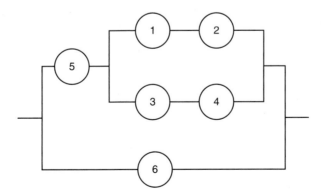

**FIGURE 10.3**    Example of reducible structure.

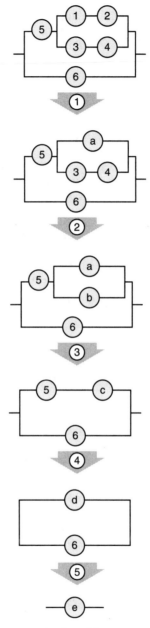

**FIGURE 10.4**   Step-by-step reduction of the structure depicted in Figure 10.3.

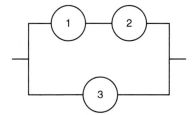

**FIGURE 10.5**    Structure of pipeline considered in Example 10.1.

**FIGURE 10.6**    Structure for Step 1.

Let the system consist of identical units with pipe throughput distribution of each unit equal to

$$p_{k1} = \Pr\{v = 100\} = 0.8,$$
$$p_{k2} = \Pr\{v = 90\} = 0.15,$$
$$p_{k3} = \Pr\{v = 80\} = 0.05,$$
$$p_{k4} = \Pr\{\text{complete failure}\} = 0.$$

*Remark.* Pipe run throughout is taken in some conditional units.

- *Step 1.* Consider first the upper branch of the pipeline consisting of a series connection of units 1 and 2 (Figure 10.6).
  For this step, we have to use the operator $\overset{\otimes}{\text{min}}$, since

$$f^{(v)}_{\text{SERIES}}(v_k, v_j) = \min(v_k, v_j).$$

Results of calculations are presented in the table below.

|  |  |  | Unit 1 | | | |
|---|---|---|---|---|---|---|
|  |  |  | State 1 | State 2 | State 3 | State 4 |
|  |  |  | (0.8; 100) | (0.15; 90) | (0.05; 80) | (0; 0) |
| Unit 2 | State 1 | (0.8; 100) | $0.8 \cdot 0.8 = 0.64$ | $0.15 \cdot 0.8 = 0.12$ | $0.05 \cdot 0.8 = 0.04$ | (0; 0) |
|  |  |  | $\min(100, 100) = 100$ | $\min(90, 100) = 90$ | $\min(80, 100) = 80$ |  |

| | | | | | |
|---|---|---|---|---|---|
| State 2 | (0.15; 90) | 0.8·0.15 = 0.12 | 0.15·0.15 = 0.0225 | 0.05·0.15 = 0.0075 | (0; 0) |
| | | min(100, 90) = 90 | min(90, 90) = 90 | min(80, 90) = 80 | |
| State 3 | (0.05; 80) | 0.8·0.05 = 0.04 | 0.15·0.05 = 0.0075 | 0.05·0.05 = 0.0025 | (0; 0) |
| | | min(100, 80) = 80 | min(90, 80) = 80 | min(80, 80) = 80 | |
| State 4 | (0; 0) | (0; 0) | (0; 0) | (0; 0) | (0; 0) |

In result, one obtains a new equivalent unit with the following distribution of throughputs:

$$P_1 = \Pr\{v = 100\} = 0.64,$$
$$P_2 = \Pr\{v = 90\} = 0.12 + 0.12 + 0.0225 = 0.2625,$$
$$P_3 = \Pr\{v = 80\} = 0.04 + 0.0075 + 0.04 + 0.0075 + 0.0025 = 0.0975,$$
$$P_4 = \Pr\{v = 0\} = 0.$$

- *Step 2.* This new equivalent unit has to be combined in parallel with the third unit (Figure 10.7).

  In this case, we have to use the operator $\overset{\otimes}{+}$ because

$$f^{(v)}_{\text{PARALLEL}}(v_k, \cdot v_j) = v_k + v_j.$$

Results of calculations are presented in the table below.

| | | | Equivalent Unit | | | |
|---|---|---|---|---|---|---|
| | | | State 1<br>(0.64; 100) | State 2<br>(0.2625; 90) | State 3<br>(0.0975; 80) | State 4<br>(0; 0) |
| Unit 3 | State 1 | (0.8; 100) | 0.64·0.8 = 0.512 | 0.2625·0.8 = 0.21 | 0.0975·0.8 = 0.078 | 0 |
| | | | 100 + 100 = 200 | 90 + 100 = 190 | 80 + 100 = 180 | 100 |
| | State 2 | (0.15; 90) | 0.64·0.15 = 0.096 | 0.2625·0.15 ≈ 0.039 | 0.0975·0.15 ≈ 0.015 | 0 |
| | | | 100 + 90 = 190 | 90 + 90 = 180 | 80 + 90 = 170 | 90 |
| | State 3 | (0.05; 80) | 0.64·0.05 = 0.032 | 0.2625·0.05 ≈ 0.013 | 0.0975·0.05 ≈ 0.005 | 0 |
| | | | 100 + 80 = 180 | 90 + 80 = 170 | 80 + 80 = 160 | 80 |
| | State 4 | (0; 0) | 0 | 0 | 0 | 0 |
| | | | 100 | 90 | 80 | 0 |

**FIGURE 10.7**   Structure for Step 2.

Resulting pipeline throughput distribution is

$$P_1 = \Pr\{v = 200\} = 0.512,$$
$$P_2 = \Pr\{v = 190\} = 0.096 + 0.21 = 0.306,$$
$$P_3 = \Pr\{v = 180\} = 0.032 + 0.039 + 0.078 = 0.149,$$
$$P_4 = \Pr\{v = 170\} = 0.013 + 0.015 = 0.028,$$
$$P_5 = \Pr\{v = 160\} = 0.005.$$

These results allow calculating the average power of the pipeline:

$$
\begin{aligned}
V_{\text{average}} &= 0.512 \cdot 200 + 0.306 \cdot 190 + 0.149 \cdot 180 + 0.028 \cdot 170 \\
&\quad + 0.005 \cdot 160 \\
&\approx 192.9.
\end{aligned}
$$

One can also find the PFFO of this system by some chosen criterion of failure. For instance, if a failure criterion is $V < 190$, then PFFO is equal to

$$P = \Pr\{V \geq 190\} = 0.512 + 0.306 = 0.818.$$

**Example 10.2**

Consider a pipeline with a structure depicted in Figure 10.8.

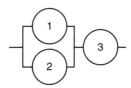

**FIGURE 10.8**   System structure for Example 10.2.

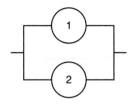

**FIGURE 10.9**    Structure for Step 1.

Let us use the following throughput distributions for the system units:

$$p_{11} = p_{21} = \Pr\{v = 100\} = 0.8,$$
$$p_{12} = p_{22} = \Pr\{v = 90\} = 0.15,$$
$$p_{13} = p_{23} = \Pr\{v = 80\} = 0.05,$$
$$p_{14} = p_{24} = \Pr\{\text{complete failure}\} \approx 0,$$

and

$$p_{31} = \Pr\{v = 200\} = 0.8,$$
$$p_{32} = \Pr\{v = 180\} = 0.15,$$
$$p_{33} = \Pr\{v = 160\} = 0.05,$$
$$p_{34} = \Pr\{\text{complete failure}\} \approx 0.$$

- *Step 1.* At the beginning consider two units 1 and 2 connected in parallel (Figure 10.9).

  In this case, one has to use operator $\overset{\otimes}{+}$, since

$$f^{(v)}_{\text{PARALLEL}}(v_k, v_j) = v_k + v_j.$$

Calculation results are presented in the table below.

|  |  |  | Unit 1 | | | |
|---|---|---|---|---|---|---|
|  |  |  | State 1 | State 2 | State 3 | State 4 |
|  |  |  | (0.8; 100) | (0.15; 90) | (0.05; 80) | |
| Unit 2 | State 1 | (0.8; 100) | $0.8 \cdot 0.8 = 0.64$ | $0.15 \cdot 0.8 = 0.12$ | $0.05 \cdot 0.8 = 0.04$ | 0 |
|  |  |  | $100 + 100 = 200$ | $90 + 100 = 190$ | $80 + 100 = 180$ | 100 |
|  | State 2 | (0.15; 90) | $0.8 \cdot 0.15 = 0.12$ | $0.15 \cdot 0.15 = 0.0225$ | $0.05 \cdot 0.15 = 0.0075$ | 0 |
|  |  |  | $100 + 90 = 190$ | $90 + 90 = 180$ | $80 + 90 = 170$ | 90 |

| | | | | | |
|---|---|---|---|---|---|
| State 3 | (0.05; 80) | $0.8 \cdot 0.05 = 0.04$ | $0.15 \cdot 0.05 = 0.0075$ | $0.05 \cdot 0.05 = 0.0025$ | 9 |
| | | $100 + 80 = 180$ | $90 + 80 = 170$ | $80 + 80 = 160$ | 80 |
| State 4 | (0; 0) | 0 | 0 | 0 | |
| | | 100 | 90 | 80 | |

In result, one obtains a new equivalent unit with the following distribution of throughputs:

$$P_1 = \Pr\{v = 200\} = 0.64,$$
$$P_2 = \Pr\{v = 190\} = 0.12 + 0.12 = 0.24,$$
$$P_3 = \Pr\{v = 180\} = 0.04 + 0.0225 + 0.04 = 0.1025,$$
$$P_4 = \Pr\{v = 170\} = 0.0075 + 0.0075 = 0.015,$$
$$P_5 = \Pr\{v = 160\} = 0.0025,$$
$$P_6 = \Pr\{v = 0\} = 0.$$

- *Step 2.* At this step, the obtained above equivalent unit has to be combined with the third unit using the operator $\underset{\min}{\otimes}$ (Figure 10.10). Calculations for the entire system are presented in the table below.

| | | Unit 3 | | | |
|---|---|---|---|---|---|
| | | State 1 (0.8; 200) | State 2 (0.15; 180) | State 3 (0.05; 160) | State 4 (0; 0) |
| Equivalent unit | State 1 (0.64; 200) | $0.8 \cdot 0.64 = 0.512$ min(200, 200) | $0.15 \cdot 0.64 = 0.096$ min(180, 200) = 180 | $0.05 \cdot 0.64 = 0.032$ min(160, 200) = 160 | 0 min(0, 200) = 0 |
| | State 2 (0.24; 190) | $0.8 \cdot 0.24 = 0.192$ min(200, 190) = 190 | $0.15 \cdot 0.24 = 0.036$ min(180, 190) = 180 | $0.05 \cdot 0.24 = 0.012$ min(160, 190) = 160 | 0 min(0, 190) = 0 |
| | State 3 (0.103; 180) | $0.8 \cdot 0.103 \approx 0.082$ min(200, 180) = 180 | $0.15 \cdot 0.103 \approx 0.015$ min(180, 180) = 180 | $0.05 \cdot 0.103 \approx 0.005$ min(160, 180) = 160 | 0 min(0, 180) = 0 |
| | State 4 (0.015; 170) | $0.8 \cdot 0.015 = 0.012$ min(200, 170) = 170 | $0.15 \cdot 0.015 \approx 0.002$ min(180, 170) = 170 | $0.05 \cdot 0.015 \approx 0$ min(160, 170) = 160 | 0 min(0, 170) = 0 |
| | State 5 (0.003; 160) | $0.8 \cdot 0.003 \approx 0.002$ min(200, 160) = 160 | $0.15 \cdot 0.003 \approx 0$ min(180, 160) = 160 | $0.05 \cdot 0.003 \approx 0$ min(160, 160) = 160 | min(0, 160) = 0 |
| | State 6 (0; 0) | 0 min(200, 0) = 0 | 0 min(180, 0) = 0 | 0 min(160, 0) = 0 | min(0, 0) = 0 |

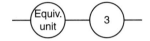

**FIGURE 10.10** Structure for Step 2.

These results allow calculating the expected throughput of the pipeline:

$$E\{V\} = 0.512 \cdot 200 + 0.192 \cdot 190 + (0.082 + 0.096 + 0.036 + 0.015) \cdot 180$$
$$+ (0.012 + 0.002) \cdot 170 + (0.002 + 0.032 + 0.012 + 0.005) \cdot 160$$
$$= 0.512 \cdot 200 + 0.192 \cdot 190 + 0.229 \cdot 180 + 0.014 \cdot 170 + 0.051 \cdot 160$$
$$\approx 190.6.$$

One can also find the PFFO of this system by some chosen criterion of failure. For instance, if a failure criterion is $V < 180$, then PFFO is equal to

$$P = \Pr\{V \geq 180\} = 0.512 + 0.192 + 0.229 = 0.933.$$

We restrict ourselves to these two numerical examples, since from methodological viewpoint considering additional structures or different physical objects brings nothing new.

## 10.7 CONCLUSION

We limit ourselves to a few examples that demonstrate main ideas of using UGF in reliability analysis. Of course, the reader can find objects of different physical nature and compile for them corresponding composition operators. Note that UGF methodology is not limited by reliability problems. It can be successfully used for other multidimensional "generalized convolutions."

## 10.8 BRIEF HISTORICAL OVERVIEW AND RELATED SOURCES

Here we offer only papers and highly related books to the subject of this chapter. List of general monographs and textbooks, which can include this topic, is given in the main bibliography at the end of the book. All references are ordered chronologically, since it gives to the reader additional information about the history of the considered subject.

First publications on multistate systems appeared in the end of 1970s and beginning of 1980s. The theory of multistate system reliability is now developed in a powerful branch of the modern reliability theory. Here we made an attempt to demonstrate the place and role of UGF, which first appeared in Ushakov (1986).

Now this new direction is flourishing mostly due to intensive and quite productive research by G. Levitin and A. Lisnianski.

Bibliography below is given in chronological–alphabetical order for better exposition of historical background of the subject.

## BIBLIOGRAPHY

Barlow, R. and A. Wu (1978) Coherent systems with multi-state components. *Math. Oper. Res.*, Vol. 3.

El-Neweihi, E., F. Proschan, and J. Seturaman (1978) Multistate coherent systems. *J. Appl. Probab.*, Vol. 15.

Griffith, W. (1980) Multistate reliability models. *J. Appl. Probab.*, Vol. 17.

Fardis, M. N. and C. A. Cornell (1981) Analysis of coherent multistate systems. *IEEE Trans. Reliab.*, Vol. 30.

Fardis, M. and C. Cornell (1982) Multistate reliability analysis. *Nucl. Eng. Des.*, Vol. 60.

Hudson, J. C. and K. Kapur (1982) Reliability theory for multistate systems with multistate components. *Microelectron. Reliab.*, Vol. 22, No. 1.

Hudson. J. C. and K. C. Kapur (1983) Reliability analysis of multistate systems with multistate components. *Trans. Inst. Ind. Eng.*, Vol. 15.

Ebrahimi N. (1984) Multistate reliability models. *Naval Res. Log.*, Vol. 31.

Ohio, F. and T. Nishida (1984) On multistate coherent systems. *IEEE Trans. Reliab.*, Vol. 33.

Aven, T. (1985) Reliability evaluation of multi-state systems with multi-state components. *IEEE Trans. Reliab.*, Vol. 34, No. 5.

Xue J. (1985a) On multistate system analysis. *IEEE Trans. Reliab.*, Vol. 34.

Xue, J. (1985b) New approach for multistate system analysis. *Reliab. Eng.*, Vol. 10.

Natvig B., S. Sormo, A. Holen, and G. Hogasen (1986) Multistate reliability theory: a case study. *Adv. Appl. Probab.*, Vol. 18.

Ushakov, I. (1986) A universal generating function. *Sov. J. Comput. Syst. Sci.*, Vol. 24, No. 5.

Ushakov, I. (1988) Reliability analysis of multi-state systems by means of modified generating function. *J. Inform. Process Cybern.*, Vol. 24, No. 3.

Aven, T. (1993) On performance measures for multistate monotone systems. *Reliab. Eng. Syst. Saf.*, Vol. 41.

Gnedenko, B. V. and I. A. Ushakov (1995) *Probabilistic Reliability Engineering*. Wiley, New York.

Ushakov, I. A. (1998) An object oriented approach to generalized generating function. Proceedings of the ECCO-XI Conference (European Chapter on Combinatorial Optimization), Copenhagen.

Levitin, G. (1999) A universal generating function approach for analysis of multi-state systems with dependent elements. *Reliab. Eng. Syst. Saf.*, Vol. 84, No. 3.

Levitin, G. and A. Lisnianski (1999) Importance and sensitivity analysis of multi-state systems using universal generating functions method. *Reliab. Eng. Syst. Saf.*, Vol. 65.

Chakravarty, S. and I. Ushakov (2000) Effectiveness analysis of Globalstar[TM] gateways. Proceedings of Second International Conference on Mathematical Methods in Reliability (MMR'2000), Vol. 1, Bordeaux, France.

Ushakov, I. (2000) The method of generalized generating sequences. *Eur. J. Oper. Res.*, Vol. 125 No. 2.

Chakravarty, S. and I. Ushakov (2002) Reliability measure based on average loss of capacity. *Int. Trans. Oper. Res.*, Vol. 9, No. 2.

Rykov, V. and B. Dimitrov (2002) On multi-state reliability systems. *Inform. Process.*, Vol. 2, No. 2.

Levitin, G., A. Lisnianski, and I. Ushakov (2003) Reliability of multi-state systems: a historical overview. In: *Mathematical and Statistical Methods in Reliability*. World Scientific.

Lisnianski, A. and G. Levitin (2003) *Multi-State System Reliability. Assessment, Optimization, Applications*. World Scientific.

Levitin, G. (2004) A universal generating function approach for the analysis of multi-state systems with dependent elements. *Reliab. Eng. Syst. Saf.*, Vol. 84, No. 3.

Levitin, G. (2005) *The Universal Generating Function in Reliability Analysis and Optimization.* Springer.

Ding, Y. and A. Lisnianski (2008) Fuzzy universal generating functions for multi-state system reliability assessment. *Fuzzy Sets Syst.*, Vol. 159, No. 3.

# APPENDIX A

# MAIN DISTRIBUTIONS RELATED TO RELIABILITY THEORY

Since each of the reliability indices represents one or another characteristic of probability distribution, we give brief information about main distributions related to the subject.

## A.1 DISCRETE DISTRIBUTIONS

### A.1.1 Degenerate Distribution

In a sense, it is a distribution of "nonrandom random variable": it is the distribution of a constant value. In reliability theory, a constant is used, for instance, for description of switching time to a redundant unit or duration of monitoring tests.

The degenerate distribution is localized at a point $T$ on the real line. The cumulative distribution function of the degenerate distribution concentrated at point $T$ is

*Probabilistic Reliability Models*, First Edition. Igor Ushakov.
© 2012 John Wiley & Sons, Inc. Published 2012 by John Wiley & Sons, Inc.

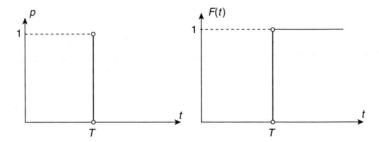

**FIGURE A.1** Probability mass function and distribution function of degenerate distribution.

$$F(t) = \begin{cases} 0, & \text{if } t < T, \\ 1, & \text{otherwise.} \end{cases} \quad (A.1)$$

Its probability mass function and distribution function are depicted in Figure A.1.

### A.1.2 Bernoulli Distribution

Let two mutually exclusive events are possible: success, which occurs with probability $p$, and failure, which occurs with probability $q = 1 - p$. Assign conditionally value 1 to success and value 0 to failure and introduce a random variable $X$ that is called a *Bernoulli*[1] *random variable*:

$$X = \begin{cases} 1, & \text{if success has occurred,} \\ 0, & \text{otherwise.} \end{cases}$$

Distribution of the Bernoulli r.v. is called the Bernoulli distribution:

$$\Pr\{x = X\} = \begin{cases} p, & \text{if } X = 1, \\ q, & \text{if } X = 0, \end{cases} \quad (A.2)$$

or in more compact form:

$$\Pr\{x = X\} = p^X q^{X-1}. \quad (A.3)$$

Thus, the Bernoulli r.v. is a special case of degenerate r.v. when $T = 1$.

---

[1] Jacob Bernoulli (1654–1705) was a Swiss mathematician, one of the many prominent mathematicians in the Bernoulli family.

The mathematical expectation (the mean) of a Bernoulli random variable $X$ is

$$E\{X\} = 1 \cdot p + 0 \cdot q = p, \tag{A.4}$$

and its variance is

$$\sigma^2\{X\} = pq. \tag{A.5}$$

### A.1.3  Binomial Distribution

If one observes a series of $n$ Bernoulli r.v.'s, the number of successes (and, respectively, failures) is random. The distribution of this r.v. is called binomial distribution. Binomial r.v. is the sum of Bernoulli r.v.'s, that is,

$$X = X_1 + X_2 + \cdots + X_n = \sum_{k=1}^{n} X_k, \tag{A.6}$$

where $X_k$ is Bernoulli r.v.

The probability of occurrence of exactly $k$ successes is

$$P(n; k) = \binom{n}{k} p^k q^{n-k}. \tag{A.7}$$

Distribution function in this case is written as

$$F(n; m) = \sum_{k=0}^{m} \binom{n}{k} p^k q^{n-k}. \tag{A.8}$$

An example of the binomial distribution is given in Figure A.2.

Using (A.4) and applying the theorem about expectation of the sum of r.v.'s, one can immediately write

$$E\{X_1 + X_2 + \cdots + X_n\} = nE\{X_1\} = np. \tag{A.9}$$

Using (A.5) and the formula for the sum of variances, one gets

$$\sigma^2 = npq. \tag{A.10}$$

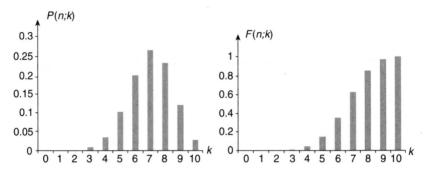

**FIGURE A.2**   Mass function, $P(n; k)$, and cumulative function, $F(n; k)$, for binomial distribution with parameters $p = 0.7$ and $n = 10$.

Let us underline an important property of binomial distribution: if one performs two series of Bernoulli tests of $n_1$ and $n_2$ trials and probability of success in both cases is $p$, then it is equivalent to a single test with the total number of trials equal to $n = n_1 + n_2$. In other words, the convolution of two Binomial distribution functions produces the binomial distribution with a new parameter $n$.

### A.1.4   Poisson Distribution

This distribution is often used in various mathematical reliability models. For instance, it describes distribution of events of the Poisson process in a fixed time interval. The "physical" sense of this distribution can be demonstrated by the following example. Consider the binomial distribution for the case when the number of Bernoulli trials, $n$, is extremely large and, at the same time, value of $p$ is very close to 1. In this case, numerical calculation of probability (A.7) presents a definite complexity: one needs to sum up a huge number of summands, each of which is a product of large value of binomial coefficient and very small values of probabilities. In this case, it is reasonable to use a limit passage for $n \to \infty$, $q \to 0$, and $nq = \text{constant}$:

$$\lim_{n \to \infty} \binom{n}{k} q^k p^{n-k} = \frac{q^k}{k!} \lim_{n \to \infty} [n \cdot (n-1) \cdots (n-k+1)](1-q)^{n-k}.$$

$$(A.11)$$

Since $n \gg k$, $\lim_{n \to \infty} [n \cdot (n-1) \cdots (n-k+1)] = n^k$ and $\lim_{n \to \infty} (1-q)^{n-k} = \exp(-nq)$. As a result, one gets

$$\lim_{n \to \infty} \binom{n}{k} q^k p^{n-k} = \frac{(nq)^k}{k!} \exp(-nq). \qquad (A.12)$$

Formula (A.12) is the corollary of the fundamental Poisson theorem.

Formula (A.12) can be used as approximation even for relatively small $n$. In Figure A.3, four particular cases for $n = 20$ are depicted: B-1 is binomial distribution with $q = 0.1$, P-1 is corresponding Poisson distribution with $\Lambda = 0.1 \cdot 20 = 0.2$, B-2 is binomial distribution with $q = 0.25$, and P-2 is corresponding Poisson distribution with $\Lambda = 0.25 \cdot 20 = 5$.

Now consider that during time interval $[0, t]$ random failures can occur with constant rate $\lambda$. Divide interval $[0, t]$ into small subintervals $\Delta$. Then in each interval may occur $\lambda\Delta + o(\Delta)$ failures, where $o(\Delta)$ is a value of higher order of smallness. In other words, we consider $n = t/\Delta$ Bernoulli trials. So, we can replace value $nq$ in this case for $\Lambda = \lambda t$. It means that (A.12) can be rewritten as

$$p_k(\Lambda) = \frac{\Lambda^k}{k!} e^{-\Lambda}. \qquad (A.13)$$

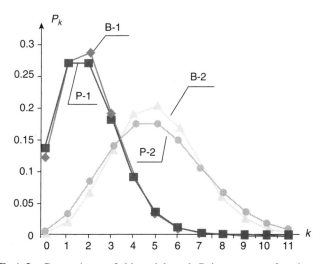

**FIGURE A.3** Comparison of binomial and Poisson mass functions for two particular cases.

Cumulative function in this case is

$$P_{j \leq k}(\Lambda) = \sum_{j=0}^{k} \frac{\Lambda^j}{j!} e^{-\Lambda}. \tag{A.14}$$

Examples of mass function and cumulative function of the Poisson distribution are given in Figure A.4a and b, respectively.

The mathematical expectation for Poisson distribution is defined in the usual way:

$$E\{k\} = \sum_{k=0}^{\infty} k \frac{(\lambda t)^k}{k!} e^{-\lambda t} = \lambda t \sum_{k=1}^{\infty} \frac{(\lambda t)^k}{(k-1)!} e^{-\lambda t} = \lambda t\, e^{-\lambda t} \sum_{k=0}^{\infty} \frac{(\lambda t)^k}{k!} = \lambda t. \tag{A.15}$$

The variance of this distribution is equal to its mean:

$$\sigma^2 = \lambda t. \tag{A.16}$$

### A.1.5    Geometric Distribution

Consider a series of Bernoulli trials. Denote the number of consecutive successes until first failure occurrence by $X$. Distribution of such r.v. is called geometric. Its mass function is expressed as

$$\Pr\{X = k\} = p^x q. \tag{A.17}$$

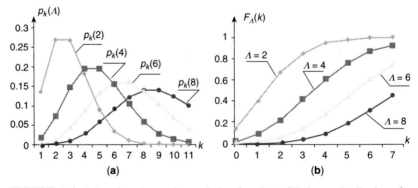

**FIGURE A.4**   Mass function and cumulative function of Poisson distributions for parameter $\Lambda$ equal to 2, 4, 6, and 8, respectively.

Cumulative function of geometric distribution has the form

$$\Pr\{X \le x\} = q\sum_{k=0}^{x} p^k = 1 - q\sum_{k=x+1}^{\infty} p^k = 1 - qp^{x+1}\sum_{k=0}^{\infty} p^k$$

$$= 1 - \frac{qp^{x+1}}{1-p} = 1 - p^{x+1}. \qquad (A.18)$$

Note that (A.18) can be obtained in different ways on the basis of simple arguments. Probability $P\{X > x\} = p^{x+1}$ is the probability that $x+1$ successes occur in a row. So, complementary probability is $P\{X \le x\} = 1 - p^{x+1}$.

The mathematical expectation of r.v. $X$ is found as usual:

$$E\{x\} = \sum_{x=0}^{\infty} xp^x q = pq \cdot \frac{d}{dp}\sum_{x=0}^{\infty} p^x = pq \cdot \frac{d}{dp}\left(\frac{1}{1-p}\right) = \frac{p}{q}. \quad (A.19)$$

By the way, the sum $\sum_{x=0}^{\infty} p^x$ can be found in another way. Denote $y = 1 + p + p^2 + \cdots$. For this infinite sum, the following equality is true: $1 + p + p^2 + \cdots = 1 + py$. From here $y = 1/(1-p) = 1/q$ if geometric series converges.

## A.2   CONTINUOUS DISTRIBUTIONS

### A.2.1   Intensity Function

For continuous distribution, there is an important additional characteristic that is often used in reliability theory. This is intensity function, $\lambda(t)$, defined as conditional density function at moment $t$ under condition that the considered r.v. is larger than $t$, that is,

$$\lambda(t) = \frac{1}{P(t)} \cdot \frac{dF(t)}{dt}. \qquad (A.20)$$

One can rewrite (A.20) as

$$\lambda(t) = -\frac{1}{P(t)} \cdot \frac{dP(t)}{dt} = \frac{d}{dt}\ln P(t). \qquad (A.21)$$

For (A.21) follows that for any distribution, the probability of failure-free operation can be written in the form

$$P(t) = \exp\left( - \int_0^t \lambda(t)dt \right).$$
(A.22)

Expression (A.22) leads beginners in reliability engineering to confusion: they call arbitrary d.f. presented in such a form "exponential distribution." Indeed, for an exponential distribution always $\lambda(t) = \text{constant}$.

For the probability that residual time to failure is larger than $t$ under condition that an object has already worked time $x$, one can write

$$P(t|x) = \exp\left( - \int_x^{t+x} \lambda(t)dt \right).$$
(A.23)

## A.2.2  Continuous Uniform Distribution

Continuous uniform distribution (or rectangular distribution) is defined on restricted interval $[x, y]$ with density function

$$f(t) = \frac{1}{y - x}$$
(A.24)

and distribution function

$$F(t) = \begin{cases} 0, & \text{if } t \leq x, \\ \dfrac{t - x}{y - x}, & \text{if } x < t \leq y, \\ 1, & \text{if } t > y. \end{cases}$$
(A.25)

Functions (A.24) and (A.25) are depicted in Figure A.5.
The mean of this distribution is equal to

$$E\{X\} = \frac{y - x}{2},$$
(A.26)

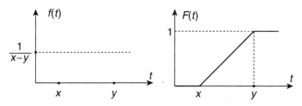

**FIGURE A.5**   Density and distribution function of uniform distribution.

and variance is equal to

$$\sigma^2 = \frac{(y-x)^2}{12}.$$  (A.27)

This distribution is widely used for Monte Carlo modeling.

### A.2.3 Exponential Distribution

Consider geometric distribution with the probability of success that is very close to 1, or, conversely, the probability of failure that is extremely small: $q = 1 - p \ll 1$. The probability that r.v. $X$, which is equal to the number of consecutive successes in such series of Bernoulli trials, is more than some fixed $n$ is equal to

$$\Pr\{X \geq n\} = p^n = (1 - q)^n.$$  (A.28)

Let each Bernoulli trial lasts time $\Delta$. Denote the failure probability as $q = \lambda \Delta$, where $\lambda$ is some constant. Assume that the number of trials, $n$, is sufficiently large. Let the total time of trials is $t = n\Delta$. Then (A.28) can be rewritten as

$$P(t) = P(X \geq n\} = (1 - \lambda \Delta)^{t/\Delta},$$  (A.29)

and after $\Delta \to 0$, one gets

$$P(t) = \exp(-\lambda t).$$  (A.30)

Formula (A.30) gives a function complementary to the distribution function, so

$$F(t) = 1 - \exp(-\lambda t).$$  (A.31)

This is exponential distribution function with parameter $\lambda$.

Thus, in a sense, exponential d.f. is a limit distribution for the geometric one. Density function for this distribution is found as follows:

$$f(t) = \frac{d}{dt}[1 - \exp(-\lambda t)] = \lambda \exp(-\lambda t). \tag{A.32}$$

The mathematical expectation is

$$E\{X\} = \int_0^\infty t\lambda \exp(-\lambda t)dt = \frac{1}{\lambda}. \tag{A.33}$$

and variance is

$$\sigma = \frac{1}{\lambda^2}. \tag{A.34}$$

It is clear that exponential distribution, as well as the geometric one, possesses the Markov property, that is,

$$P(x, t + x | x > x) = P(0, t) = \exp(-\lambda t). \tag{A.35}$$

In reliability terms, it means that an object that is in operational state at some moment $t$ is in operational state by its reliability properties that are equivalent to an absolutely new one. Of course, this assumption should be always taken into account before using exponential models for practical purposes: not always such assumption is adequate to a real technical object.

Intensity function, $\lambda(t)$, for exponential function is constant:

$$\frac{1}{P(t)} \cdot \frac{dF(t)}{dt} = \frac{1}{e^{-\lambda t}} \cdot \lambda\, e^{-\lambda t} = \lambda. \tag{A.36}$$

### A.2.4 Erlang Distribution

Erlang distribution is a convolution of $n$ identical exponential distributions; that is, Erlang r.v. is a sum of $n$ i.i.d. exponential r.v.'s. Erlang d.f. represents a particular case of gamma distributions with an integer shape parameter.

Erlang density function is

$$f(t) = \lambda \frac{(\lambda t)^{n-1}}{(n-1)!} \cdot \exp(-\lambda t), \quad t \geq 0. \tag{A.37}$$

The mean and variance of this distribution are equal to $n/\lambda$ and $n/\lambda^2$, respectively.

### A.2.5   Hyperexponential Distribution

This distribution appears in some reliability models. Hyperexponential distribution is a weighed sum of exponential d.f.'s and is defined as

$$F(t) = 1 - \sum_{k=1}^{n} p_k \exp(-\lambda_k t), \tag{A.38}$$

where $\sum_{k=1}^{n} p_k = 1$ and all $p_k > 0$.

An example of hyperexponential distribution is presented in Figure A.6.

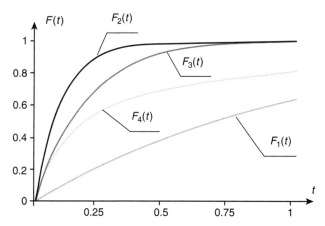

**FIGURE  A.6**   Exponential   functions   $F_1(t) = e^{-t}$   and   $F_2(t) = e^{-10t}$,   and   hyperexponential functions $F_3(t) = 0.5F_1(t) + 0.5F_2(t)$. For comparison, function $F_4(t) = e^{-5.5t}$ with parameter $\lambda = 0.5(1 + 10)$ is also presented.

Obviously, the mean of this distribution is equal to

$$E\{X\} = \sum_{k=1}^{n} \frac{p_k}{\lambda_k}.$$ (A.39)

This distribution has a decreasing intensity function:

$$\lambda(t) = \frac{\sum_{k=1}^{n} p_k \lambda_k \exp(-\lambda_k t)}{\sum_{k=1}^{n} p_k \exp(-\lambda_k t)}.$$ (A.40)

Indeed, function $\lambda(t)$ is monotone and for $t = 0$ is equal to

$$\lambda(0) = \frac{\sum_{k=1}^{n} p_k \lambda_k}{\sum_{k=1}^{n} p_k} = \sum_{k=1}^{n} p_k \lambda_k,$$ (A.41)

and for $t \to \infty$, as one can see directly from (A.41), $\lim_{t \to \infty} \lambda(t) = \min_{1 \le k \le n} \lambda_k$. It is clear that $\min_{1 \le k \le n} \lambda_k < \lambda(0)$. In this case, intensity function has the form depicted in Figure A.7.

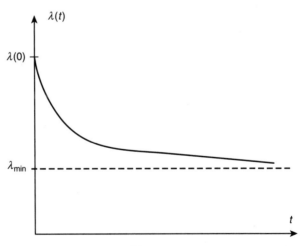

**FIGURE A.7**    Function $\lambda(t)$ for hyperexponential distribution.

## A.2.6   Normal Distribution

## A.2.7   Weibull–Gnedenko Distribution

In conclusion, let us consider Weibull–Gnedenko distribution that is widely used in applied reliability analysis. Distribution function in this case has the form

$$F(t) = \begin{cases} 1 - \exp(-(\lambda t)^\beta), & \text{for } t \geq 0, \\ 0, & \text{for } t < 0, \end{cases} \tag{A.42}$$

and density function is

$$f(t) = \begin{cases} \lambda^\beta \beta t^{\beta-1} \exp(-(\lambda t)^\beta), & \text{for } t \geq 0, \\ 0, & \text{for } t < 0. \end{cases} \tag{A.43}$$

Parameters $\lambda$ and $\beta$ are called scale and shape parameters, respectively. Examples of distribution and density functions are depicted in Figure A.8.

Weibull–Gnedenko distribution has an intensity function

$$\lambda(t) = \lambda^\beta \beta t^{\beta-1}. \tag{A.44}$$

It is increasing for $\beta > 1$ and decreasing for $\beta < 1$. Obviously, for $\beta = 1$, this distribution coincides with the exponential one. For $\beta = 1$, intensity function is linear (Figure A.9).

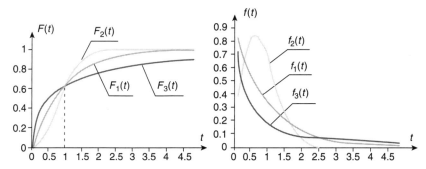

**FIGURE A.8**   Distribution, $F(t)$, and density, $f(t)$, functions. All distributions with $\lambda = 1$. Subscript "2" relates to the Weibull–Gnedenko distribution with parameter $\beta = 2$, and subscript "3" relates to the distribution with parameter $\beta = 0.5$.

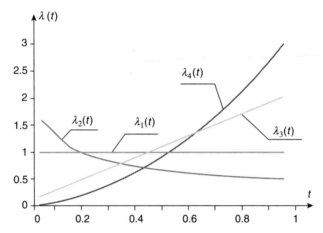

**FIGURE A.9** Examples of intensity functions for Weibull–Gnedenko distribution. All distributions with $\lambda = 1$. Subscript "1" correspond to $\beta = 1$ (exponential distribution), "2" to $\beta = 0.5$, "3" to $\beta = 2$, and "4" to $\beta = 3$.

The mean of this distribution is equal to

$$E\{X\} = \frac{1}{\lambda}\Gamma\left(1 + \frac{1}{\beta}\right),$$

(A.45)

and variance is

$$\sigma^2 = \frac{1}{\lambda^2}\left[\Gamma\left(1 + \frac{2}{\beta}\right) - \left(\Gamma\left(1 + \frac{1}{\beta}\right)\right)^2\right],$$

(A.46)

where $\Gamma(\cdot)$ is a gamma function.

# APPENDIX B

# LAPLACE TRANSFORMATION

Recall that the Laplace[1] transform (LT), $\varphi(s)$, of function $y(t)$ is defined over the positive axis as

$$\varphi(s) = \int_0^\infty y(t)e^{-st}\,dt. \tag{B.1}$$

For derivative of $y(t)$, that is, for $y'(t)$, one can write, using integration by parts,

$$\int_0^\infty y'(t)\,e^{-st}\,dt = \int_0^\infty e^{-st}\,dy(t) = y(t)\,e^{-st}\big|_0^\infty - \int_0^\infty f(t)\,d(e^{-st})$$

$$= -y(0) + s\int_0^\infty f(t)\cdot e^{-st}\,dt = -y(0) + s\varphi(s), \tag{B.2}$$

where $y(0)$ is the initial condition of the process, that is, $y(t)$ at $t=0$.

---

[1] Pierre-Simon, Marquis de Laplace (1749–1827) was a French mathematician and astronomer whose work was pivotal to the development of mathematical astronomy and statistics.

---

*Probabilistic Reliability Models*, First Edition. Igor Ushakov.
© 2012 John Wiley & Sons, Inc. Published 2012 by John Wiley & Sons, Inc.

For integral of function $y(t)$, the expression for LT can be derived by using integration by parts:

$$\int_0^\infty \left[ \int_0^t y(x)\, dx \right] \cdot e^{-st}\, dt = -\frac{1}{s} \int_0^\infty \left[ \int_0^t y(x)\, dx \right] \cdot d(e^{-st})$$

$$= \frac{1}{s} \left[ e^{-st} \int_0^t y(x)\, dx \Big| - \int_0^\infty y(t) \cdot e^{-st}\, dt \right]$$

$$= \frac{1}{s} \cdot \varphi(s).$$

$$(B.3)$$

For a system of differential equations, describing system's transit from state to state, one usually considers the following type of linear differential equations:

$$\frac{d}{dt} p_k(t) = -p_k(t) \sum_{i \in e(k)} \alpha_{ki} + \sum_{i \in E(k)} \alpha_{ik} p_i(t)$$

$$= -p_k(t) A_k + \sum_{i \in E(k)} \alpha_{ik} p_i(t), \qquad (B.4)$$

where $\alpha_{ki}$ is the passage intensity from state "$k$" to state "$i$", $e(k)$ is subset of the total set of states where the process can move at one step from state "$k$", $E(k)$ is subset of states from where the process can move at one step to state "$k$", and $A_k = \sum_{i \in e(k)} \alpha_{ki}$. If there are no absorbing states, one needs to use the initial conditions of the type $p_k(0)$, that is, $p_k(t)$ at moment $t = 0$.

Laplace transform for system (B.4) has the form

$$-p_k(0) + s\varphi_k(s) = -\phi_k(s) \sum_{i \in e(k)} \alpha_{ki} + \sum_{i \in E(k)} a_{ik}\varphi_i(s). \qquad (B.5)$$

This can be rewritten in open form as

$$-(s + A_1) \cdot \varphi_1(s) + \alpha_{21}\varphi_2(s) + \cdots + \alpha_{n1}\varphi_n(s) = p_1(0),$$
$$\alpha_{12}\varphi_1(s) - (s + A_2) \cdot \varphi_2(s) + \cdots + \alpha_{n2}\varphi_n(s) = p_2(0),$$
$$\vdots \qquad\qquad\qquad\qquad\qquad\qquad\qquad\qquad (B.6)$$
$$\alpha_{1n}\varphi_1(s) + \alpha_{2n}\varphi_2(s) + \cdots - (s + A_n) \cdot \varphi_n(s) = p_n(0).$$

If the considered Markov process has no absorbing states, equations in (B.6) are mutually dependent and in this case one has to replace any of them by the normalization equation:

$$s\varphi_1(s) + s\varphi_2(s) + \cdots + s\varphi_n(s) = 1. \qquad \text{(B.7)}$$

The same system of equations can be written in matrix form as

$$
\begin{vmatrix}
-(s+A_1) & \alpha_{21} & \cdots & \alpha_{n1} \\
\alpha_{12} & -(s+A_2) & \cdots & \alpha_{n2} \\
\vdots & \vdots & \ddots & \vdots \\
\alpha_{1n} & \alpha_{2n} & \cdots & -(s+A_n)
\end{vmatrix}
\times
\begin{vmatrix}
\varphi_1(s) \\
\varphi_2(s) \\
\vdots \\
\varphi_n(s)
\end{vmatrix}
=
\begin{vmatrix}
p_1(0) \\
p_2(0) \\
\vdots \\
p_n(0)
\end{vmatrix},
$$

$$\text{(B.8)}$$

where

$$
D =
\begin{vmatrix}
-(s+A_1) & \alpha_{21} & \cdots & \alpha_{n1} \\
\alpha_{12} & -(s+A_2) & \cdots & \alpha_{n2} \\
\vdots & \vdots & \ddots & \vdots \\
\alpha_1 & \alpha_{2n} & \cdots & -(s+A_n)
\end{vmatrix}
\qquad \text{(B.9)}
$$

is the determinant of equation system (B.6). Solution of this system can be obtained with Cramer's rule:

$$\varphi_k(s) = \frac{D_k(s)}{D(s)}. \qquad \text{(B.10)}$$

where $D_k(s)$ is the determinant in which the $k$th column is substituted by the right column of absolute terms.

For inverse Laplace transforms, one uses the following procedure:

(a) Open the numerator and denominator of fraction (B.10) and write $\varphi_k(s)$ in the form

$$\varphi_k(s) = \frac{A_0 + A_1 s + \cdots + A_n s^n}{B_0 + B_1 s + \cdots + B_{n+1} s^{n+1}}, \qquad \text{(B.11)}$$

where $A_k$ and $B_k$ are unknown coefficients to be found.

(b) Find roots of the polynomial in the denominator of fraction (B.11):

$$B_0 + B_1 s + \cdots + B_{n+1} s^{n+1} = 0. \tag{B.12}$$

Let these roots are $s_1, s_2, \ldots, s_n$. It means that

$$B_0 + B_1 s + \cdots + B_{n+1} s^{n+1} = \prod_{j=1}^{n+1} (s - s_j). \tag{B.13}$$

(c) Write $\varphi_k(s)$ as the sum of simple fractions:

$$\varphi_A(s) = \frac{\beta_1}{s - s_1} + \frac{\beta_2}{s - s_2} + \cdots + \frac{\beta_{n+1}}{s - s_{n+1}}, \tag{B.14}$$

where $\beta_k$'s are unknown coefficients.

(d) After reduction of fraction to a common denominator, write $\varphi_k(s)$ in the form

$$\varphi_k(s) = \frac{\sum_{1 \le j \le n} \beta_j \prod_{i \ne j} (s - s_i)}{\prod_{1 \le i \le n+1} (s - s_i)}. \tag{B.15}$$

(e) Open the numerator of the fraction and perform collection of terms:

$$\varphi_{E(k)}(s) = \frac{\gamma_0 + \gamma_1 s + \gamma_2 s^2 + \cdots + \gamma_n s^n}{(s - s_1)(s - s_2) \cdots (s - s_{n+1})}, \tag{B.16}$$

where $\gamma_k$'s are presented through known $\beta_j$'s and $s_j$'s.

**TABLE B.1   Most Important Laplace Transforms**

| Origin | Transform | Origin | Transform |
|---|---|---|---|
| $\alpha p_1(t) + \beta p_2(t)$ | $\alpha \varphi_1(s) + \beta \varphi_2(s)$ | $\int_0^t p_1(x) p_2(t - x)\,dx$ | $\varphi_1(s) \cdot \varphi_2(s)$ |
| $\frac{dp(t)}{dt}$ | $-p(0) + s\varphi(s)$ | $(-t)^k p(t)$ | $\varphi^{(k)}(s)$ |
| $\int_0^t p(x)\,dx + C$ | $\frac{\varphi(s)}{s} + \frac{C}{s}$ | $e^{\alpha t} p(t)$ | $\varphi(s - a)$ |
| $p(bt)$ | $\frac{1}{b} \varphi(\frac{s}{b})$, for $b > 0$ | $\frac{t^{n-1}}{(n-1)!}$ | $\frac{1}{s^n}$, for $n = 1, 2, \ldots$ |
| $p(t - b)$ | $e^{-bs} \varphi(s)$, for $b > 0$ | $\frac{t^{n-1}}{(n-1)!} \cdot e^{-\beta t}$ | $\frac{1}{(s-\beta)^n}$, for $n = 1, 2, \ldots$ |

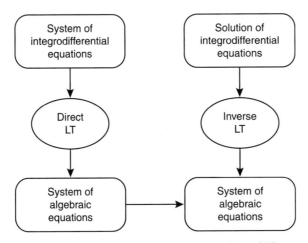

**FIGURE B.1**    Graphical explanations of the idea of LT use.

(f) Polynomials (B.11) and (B.16) are equal if and only if $A_k = \gamma_k$. From these conditions, one finds unknown coefficients $\beta_k$.

(g) After finding $\beta_k$, one applies inverse Laplace transforms to (B.14).

For practical solutions, one can use existing tables of inverse Laplace transforms. A sample of the most important Laplace transforms, frequently used in reliability analysis, is given in Table B.1.

Usage of this table is almost obvious. For instance, if your solution in Laplace transforms is $\alpha/(\alpha + s)$, it means that in the "space of normal functions" the solution is $e^{-\alpha t}$.

Thus, the main idea of LT consists in replacing integrodifferential equations by equivalent algebraic ones, which can be solved easily, and then make inverse transform for the obtained algebraic solution.

Why one needs to use LT? Explanation is simple: it makes solution of integrodifferential equations simpler. The main idea of using LT is explained in Figure B.1.

# APPENDIX C

# MARKOV PROCESSES

## C.1  GENERAL MARKOV PROCESS

From the very beginning, we would like to emphasize that a Markov model is an idealization of real processes. The main problem is not to solve the system of mathematical equations but rather to identify the real problem, to determine if the real problem and the mathematical model are an appropriate fit to each other. If, in fact, they fit, then a Markov model is very convenient.

Assume that we can construct the transition graph that sufficiently describes a system's operation. (We use below reliability terminology only for the reader's convenience.) This graph must represent a set of mutually exclusive and totally exhaustive system states with all of their possible one-step transitions. Using some criterion of system failure, all of these states can be divided into two complementary disjoint subsets, *up states* and *down states*. Necessary condition that transition from the subset of up states to the set of down states occurs is a failure of one of the operating units. Of course, if a unit is redundant, the system failure does not occur. An inverse transition may occur only if a

*Probabilistic Reliability Models*, First Edition. Igor Ushakov.
© 2012 John Wiley & Sons, Inc. Published 2012 by John Wiley & Sons, Inc.

failed unit is recovered by either a direct repair or a replacement. Let us consider a system with $n$ units. Any system state may be denoted by a binary vector $s = (s_1, \ldots, s_n)$, where $s_i$ is the state of the $i$th unit and $n$ is the number of units in the system. We set $s_i = 1$ if the unit is operational and $s_i = 0$ otherwise. If each system unit has two states, say, operational and failure, and the system consists of $n$ units, the system, in principle, can have $N = 2n$ different states. System state $(s_1 = 1, \ldots, s_i = 1, \ldots, s_n = 1)$ will play a special role in further deductions, so let us assign to this state subscript "1".

The transition from $(s_1, \ldots, s_i = 1, \ldots, s_n)$ to $(s_1, \ldots, s_i = 0, \ldots, s_n)$ means that the $i$th unit changes its state from up to down. The *transition rate* (or the *transition intensity*) for this case is equal to the $i$th unit's failure rate.

A transition from system state $(s_1, \ldots, s_i = 0, \ldots, s_n)$ to state $(s_1, \ldots, s_i = 1, \ldots, s_n)$ means that the $i$th unit being in a failed state has been recovered. The transition rate for this case is equal to the $i$th unit's repair rate. These kinds of transitions are most common. For Markov models, no more than a single failure can occur during infinitesimally small period of time.

For some reasons, sometimes one introduces absorbing states, that is, such states that process once entering the state will never leave it. The sense of these states will be explained later.

We denote transitions from state to state on transition graphs with arrows. The rates (intensities) are denoted as weights on the arrows.

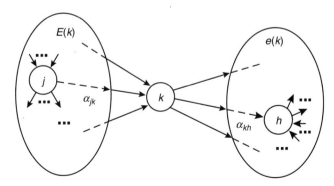

**FIGURE C.1**    Conditional presentation of state $k$ and its immediate neighbors.

After the transition graph has been constructed, it can be used as a visual aid to determine different reliability indices.

Consider a continuous Markov process with discrete states (Figure C.1). On this graph, the process can move during infinitesimally small time interval from state $k$ to one of the states of subset $e(k)$, and at the same time the process can occur in state from some subset $E(k)$. Note that a state can belong simultaneously to both subsets, that is, the process can go from one state to another back and forth.

## C.1.1 Nonstationary Availability Coefficient

This reliability index can be compiled with the help of the above transition graph. In this case, one should choose the initial state, that is, state in which system is at moment $t = 0$. For arbitrary state $k$, one can write the following formula of total probability:

$$p_k(t + \Delta) = p_k(t) \cdot \left[ 1 - \Delta \sum_{i \in e(k)} \alpha_{kj} \right] + p_j(t) \cdot \left[ \Delta \sum_{j \in E(k)} \alpha_{jk} \right], \quad \text{(C.1)}$$

where $\alpha_{ki}$ is the transition intensity from state $k$ to state $i$, $e(k)$ is subset of the total set of states where the process can move at one step from state $k$, and $E(k)$ is subset of states from where the process can move at one step to state $k$.

Indeed, the process occurs at moment $t + \Delta$ in state $k$ by two ways:

(a) At moment $t$, it is in state $k$ and does not leave this state during infinitesimally small interval $\Delta$.
(b) At moment $t$, it is in any state belonging to subset $e(k)$ and moves to state $k$ during infinitesimally small interval $\Delta$.

Equation (C.1) can be transformed into the form

$$\frac{p_k(t + \Delta) - p_k(t)}{\Delta} = -p_k(t) \sum_{i \in e(k)} \alpha_{ki} + p_j(t) \sum_{j \in E(k)} \alpha_{jk}. \quad \text{(C.2)}$$

Limiting passage for $\Delta \to 0$ leads to the following differential equation for state $k$:

$$\frac{dp_k(t)}{dt} = -p_k(t) \sum_{i \in e(k)} \alpha_{ki} + \sum_{j \in E(k)} \alpha_{jk} p_j(t), \quad \text{for } k = 1, \ldots, N, \quad (C.3)$$

with the initial condition $p_1(0) = 1$.

This system of differential equations in open form is

$$\frac{dp_1(t)}{dt} = -p_1(t) \sum_{i \in e(1)} \alpha_{1i} + \alpha_{21} p_2(t) + \cdots + \alpha_{N-1;1} p_{N-1}(t),$$

$$\frac{dp_2(t)}{dt} = \alpha_{12} p_1(t) - p_2(t) \sum_{i \in e(2)} \alpha_{2i} + \cdots + \alpha_{N-1;2} p_{N-1}(t),$$

$$\vdots$$

$$\frac{dp_{N-1}(t)}{dt} = \alpha_{1;N-1} p_1(t) + \alpha_{2;N-1} p_2(t) + \cdots - p_{N-1}(t) \sum_{i \in e(N-1)} \alpha_{N;N-1} + p_N(t),$$

$$p_1(t) + p_2(t) + \cdots + p_N(t) = 1.$$

$$(C.4)$$

On the basis of (C.4) and using Appendix B, one can write the following algebraic system of equations in terms of Laplace transforms:

$$-1 - \left( s + \sum_{i \in e(1)} \lambda_{1j} \right) \cdot \varphi_1(s) + \lambda_{21} \varphi_2(s) + \cdots + \lambda_{k1} \varphi_k(s) = 0,$$

$$\lambda_{12} \varphi_1(s) - \left( s + \sum_{i \in e(2)} \lambda_{2j} \right) \cdot \varphi_2(s) + \cdots + \lambda_{k2} \varphi_k(s) = 0,$$

$$\vdots$$

$$\lambda_{1;N-1} \varphi_1(s) + \lambda_{2;N-1} \varphi_2(s) + \cdots - \left( s + \sum_{i \in e(N-1)} \lambda_{kj} \right) \cdot \varphi_k(s) = q_m^*,$$

$$\sum_{k=1}^{N} \varphi_k(s) = \frac{1}{s}.$$

$$(C.5)$$

General form of solution of the considered equation system with the help of Laplace transforms is given in Appendix B.

### C.1.2  Probability of Failure-Free Operation

For finding the probability of a failure-free operation, absorbing states are introduced into the transition graph. They are the system's failure states (Figure C.2).

We can change the domain of summation in the previous equations in a way that is equivalent to eliminating the zero transition rates. Using the previous notation, we can immediately write for an operational state $k$ formally the same equation as (C.3).

If the transition graph has $M$ operational states (denote this set of states by $\Omega$), we can construct $M$ differential equations. (In this case, the equations are not linearly dependent. So, there is no need to use the normalization condition as one of the equations.) Equations (C.5) and the initial conditions $p_k(0), k = 1, \ldots, N$, are used to find the probability of a failure-free operation of the system.

Actually, there are two special cases that are considered in reliability theory.

1. At moment $t = 0$, the system is in state $s = (1, 1, \ldots, 1)$, that is, in the state where all units are operational, with probability 1.
2. In a stationary process, the system at arbitrary moment $t$ can be in one of its $M$ operable states with stationary probabilities $p_k$, $k = 1, \ldots, M$.

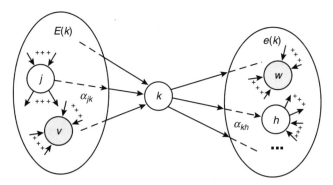

**FIGURE C.2**   Absorbing states $v$ and $w$.

Thus, systems of differential equations are the same:

$$\frac{dp_1(t)}{dt} = -p_1(t) \sum_{i \in e(1)} \alpha_{1i} + \alpha_{21}p_2(t) + \cdots + \alpha_{M;1}p_M(t),$$

$$\frac{dp_2(t)}{dt} = \alpha_{12}p_1(t) - p_2(t) \sum_{i \in e(2)} \alpha_{2i} + \cdots + \alpha_M p_M(t),$$

$$\vdots$$ 

$$\frac{dp_M(t)}{dt} = \alpha_{1;M}p_1(t) + \alpha_{2;M}p_2(t) + \cdots - p_M(t) \sum_{i \in e(M)} \alpha_{Mi}.$$

$$(C.6)$$

However, for the first case the initial condition is

$$p_0(0) = 1, \quad p_k(0) = 0, \quad k = 2, \ldots, M, \qquad (C.7)$$

and for the second case:

$$p_k(0) = p_k, \quad k = 1, \ldots, M. \qquad (C.8)$$

Thus, based on Appendix B, one can write two systems of algebraic equations in terms of Laplace transforms for the first initial condition:

$$-1 - \left(s + \sum_{j \in \Omega} \alpha_{1j}\right) \varphi_1(s) + \alpha_{21}\varphi_2(s) + \cdots + \alpha_{M1}\varphi_M(s) = 0,$$

$$\alpha_{12}\varphi_1(s) - \left(s + \sum_{j \in \Omega} \alpha_{2j}\right) \varphi_2(s) + \cdots + \alpha_{M2}\varphi_M(s) = 0,$$

$$\vdots$$

$$\lambda_{1;M}\varphi_1(s) + \lambda_{2;M}\varphi_2(s) + \cdots - \left(s + \sum_{j \in \Omega} \lambda_{M;j}\right) \varphi_M(s) = 0.$$

$$(C.9)$$

For the second initial condition, the system of algebraic equations has the form

$$-\left(s + \sum_{\forall j}\lambda_{1j}\right)\varphi_1(s) + \lambda_{21}\varphi_2(s) + \cdots + \lambda_{k1}\varphi_k(s) = q_1^*,$$

$$\lambda_{12}\varphi_1(s) - \left(s + \sum_{\forall j}\lambda_{2j}\right)\varphi_2(s) + \cdots + \lambda_{k2}\varphi_k(s) = q_2^*,$$

(C.10)

$$\vdots$$

$$\lambda_{1m}\varphi_1(s) + \lambda_{2;m}\varphi_2(s) + \cdots - \left(s + \sum_{\forall j}\lambda_{k;j}\right)\varphi_k(s) = q_m^*,$$

where $q_1^* = q_1/\sum_{k\in\Omega}q_k$.

Solution of both these systems of equations can be found with the methods described in Appendix B.

### C.1.3  Stationary Availability Coefficient

This reliability index is found on the basis of a transition graph that has only transition states. In general graph, for each state $k$ let us mark out corresponding subsets $E(k)$ and $e(k)$.

On the basis of Figure C.1, one can easily write the following balance equation for each $k$:

$$p_k \sum_{i\in e(k)} \lambda_{ki} = \sum_{j\in E(k)} \lambda_{jk}p_j.$$

(C.11)

This equation has a simple physical interpretation. Imagine that state $k$ is some reservoir with volume of liquid $p_k$ (i.e., it is proportional to the probability of time that the process stays in this state). Each liquid unit of volume flows out with total intensity $\sum_{i\in e(k)}\lambda_{ki}$ into corresponding reservoirs that belong to subset $e(k)$. In other words, the total "flow of liquid" from reservoir $k$ is equal to $p_k\sum_{i\in e(k)}\lambda_{ki}$. At the same time, liquid flows into reservoir $k$ from reservoirs belonging to subset $E(k)$. The total liquid volume flow into reservoir $k$ is equal to $\sum_{j\in E(k)}\lambda_{jk}p_j$. Our intuition hints that after a while it will be a kind of dynamic balance: the volume of liquid flowing into each reservoir will be equal to the volume of liquid flowing out.

Since the transition graph has no absorbing states, one has to take any $(n-1)$ equations of type (C.11) and add to them the so-called "normalization condition":

$$\sum_{k=1}^{n} p_k = 1. \tag{C.12}$$

In canonical form, the system of equations is

$$-p_1 \sum_{\forall j} \lambda_{1j} + \lambda_{21}p_2 + \cdots + \lambda_{n1}p_n = 0,$$

$$\lambda_{12}p_1 - p_2 \sum_{\forall j} \lambda_{2j} + \cdots + \lambda_{n2}p_n = 0,$$

$$\vdots$$

$$\lambda_{1;n-1}p_1 + \lambda_{2;n-1}p_2 + \cdots - p_{n-1} \sum_{\forall j} \lambda_{n-1;n-1} + \lambda_{n;n-1}p_n = 0,$$

$$p_1 + p_2 + \cdots + p_n = 1. \tag{C.13}$$

Solution of this system of algebraic equations is possible, for instance, with the help of Cramer's rule:

$$p_k = \frac{D_k}{D}, \tag{C.14}$$

where $D$ is the determinant of the system (C.13) and $D_k$ is the same determinant in which the $k$th column is substituted for the column of the absolute terms. Availability coefficient, $K$, is

$$K = \sum_{k \in \Omega} p_k = \frac{1}{D} \sum_{k \in \Omega} D_k, \tag{C.15}$$

where $\Omega$ is the subset of all operational states.

## C.1.4 Mean Time to Failure and Mean Time Between Failures

Recall that the mean time is defined as

$$T = \int_0^{\infty} P(t)\, dt. \tag{C.16}$$

Now note that if $\varphi(s)$ is Laplace transform for $P(t)$, then $T$ can be obtained as

$$T = \left[ \int_0^\infty P(t)e^{-st}\, dt \right]_{s=0} = \varphi(s)|_{s=0}. \qquad (C.17)$$

To find MTTF, one has to take solution of equation system (C.6) with the initial condition (C.7). For MTBF, one has to take solution of the same equation system with the initial condition (C.8).

## C.1.5  Mean Recovery Time

Finding this index is analogous to the finding of MTBF; however, in this case the sets of transitive and absorbing states have to be redefined. In this case, all operable states become absorbing and all failure states become transitive. If the total number of system states is equal to $n$, and among them there are $m$ "up" states, then for remaining $k = n - m$ states (denote the set of these states by $g$) one has to compile system of $k$ equations:

$$-\left(s + \sum_{\forall j}\lambda_{1j}\right)\varphi_1(s) + \lambda_{21}\varphi_2(s) + \cdots + \lambda_{k1}\varphi_k(s) = q_1^*,$$

$$\lambda_{12}\varphi_1(s) - \left(s + \sum_{\forall j}\lambda_{2j}\right)\varphi_2(s) + \cdots + \lambda_{k2}\varphi_k(s) = q_2^*,$$

$$\vdots \qquad\qquad\qquad\qquad\qquad\qquad\qquad\qquad (C.18)$$

$$\lambda_{1m}\varphi_1(s) + \lambda_{2;m}\varphi_2(s) + \cdots - \left(s + \sum_{\forall j}\lambda_{kj}\right)\varphi_k(s) = q_m^*,$$

where each $q_k^*$, $k \in g$, is defined as

$$q_k^* = \frac{p_j}{\sum_{j\in g}p_j}. \qquad (C.19)$$

Solution of system (C.18) can be found as

$$\tau = \sum_{j\in g}\varphi_j(0). \qquad (C.20)$$

## C.2   BIRTH–DEATH PROCESS

Birth–death process (BDP) is one of the most important special cases of the continuous-time homogenous Markov process where the states represent the current size of a population. This process has many applications in demography, biology, queuing theory, reliability engineering, and other areas.

Let us denote states by natural numbers 0, 1, 2, . . . If the process at moment $t$ is in state $k$, then during infinitesimally small time interval $\Delta$ it can with probability $\lambda_k \Delta + o(\Delta)$ proceed to state $(k+1)$, or with probability $\mu_k \Delta + o(\Delta)$ it can proceed to state $(k-1)$, or with probability $1 - (\lambda_k + \mu_k)\Delta + o(\Delta)$ it will stay in state $k$. Note that states "0" and "$n$" are so-called reflecting, that is, $\lambda_0 = 0$ and $\mu_n = 0$. Corresponding transition graph is presented in Figure C.3.

For state $k$, one can write the following equation of dynamic balance:

$$
\begin{aligned}
p_k(t + \Delta) = &\, p_{k-1}(t)[\lambda_{k-1}\Delta + o(\Delta)] \\
&+ p_k(t)[1 - (\lambda_k + \mu_k)\Delta + o(\Delta)] \\
&+ p_{k+1}(t)[\mu_k \Delta + o(\Delta)].
\end{aligned}
\tag{C.21}
$$

From (C.21) follows

$$
\frac{p_k(t + \Delta) - p_k(t)}{\Delta} = \lambda_{k-1}p_{k-1}(t) - p_k(t)(\lambda_k + \mu_k) + \mu_k p_{k+1}(t).
\tag{C.22}
$$

This gives after limiting passage for $\Delta \to 0$ the following differential equation:

$$
\frac{d}{dt}p_k(t) = \lambda_{k-1}p_{k-1}(t) - p_k(t)(\lambda_k + \mu_k) + \mu_k p_{k+1}(t).
\tag{C.23}
$$

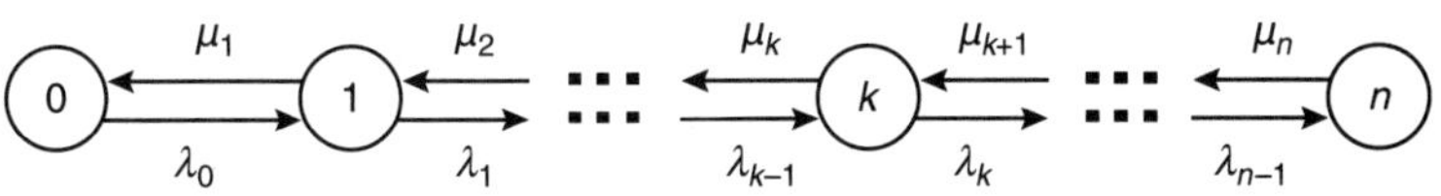

**FIGURE C.3**   Transition graph for the birth–death process.

In an analogous way, one can write equations for states "0" and "$n$":

$$\frac{d}{dt}p_0(t) = -\lambda_0 p_0(t) + \mu_1 p_1(t) \qquad (C.24)$$

and

$$\frac{d}{dt}p_n(t) = \lambda_{n-1}p_{n-1}(t) - \mu_n p_n(t). \qquad (C.25)$$

In result, we have the following system of differential equations:

$$\frac{d}{dt}p_0(t) = -\lambda_0 p_0(t) + \mu_1 p_1(t),$$

$$\frac{d}{dt}p_1(t) = \lambda_0 p_0(t) - p_1(t)(\lambda_1 + \mu_1) + \mu_2 p_2(t),$$

$$\vdots \qquad (C.26)$$

$$\frac{d}{dt}p_k(t) = \lambda_{k-1}p_{k-1}(t) - p_k(t)(\lambda_k + \mu_k) + \mu_k p_{k+1}(t),$$

$$\vdots$$

$$\frac{d}{dt}p_n(t) = \lambda_{n-1}p_{n-1}(t) - \mu_n p_n(t).$$

The initial condition in most reliability applications is taken in the form $p_0(0) = 1$.

Usually, one is interested in stationary probabilities, when $p_k(\infty) \to p_k$. It means that all $dp_k(t)/dt \to 0$ with $t \to \infty$. In this case, the system (C.26) transforms into the system of algebraic equations:

$$0 = -\lambda_0 p_0 + \mu_1 p_1,$$
$$0 = \lambda_0 p_0 - (\lambda_1 + \mu_1)p_1 + \mu_2 p_2,$$
$$\vdots$$
$$0 = \lambda_{k-1}p_{k-1} - (\lambda_k + \mu_k)p_k + \mu_k p_{k+1}, \qquad (C.27)$$
$$\vdots$$
$$0 = \lambda_{n-1}p_{n-1} - \mu_n p_n.$$

In addition, one has to use equation of total probability:

$$\sum_{k=0}^{n} p_k = 1. \tag{C.28}$$

Actually, the equations of balance for the birth–death process, written for "cuts" of the transition graph, are more convenient (Figure C.4).

Indeed, balance means that flows back and forth through a cut are equal. In this case, the system of balance equations has a very convenient form:

$$\begin{aligned}
\lambda_0 p_0 &= \mu_1 p_1, \\
\lambda_1 p_1 &= \mu_2 p_2, \\
&\ \ \vdots \\
\lambda_{k-1} p_{k-1} &= \mu_k p_k, \\
&\ \ \vdots \\
\lambda_{n-1} p_{n-1} &= \mu_n p_n.
\end{aligned} \tag{C.29}$$

Introducing $\rho_k = \lambda_k / \mu_k$, one gets the same system (C.29) in the form

$$\begin{aligned}
p_1 &= \rho_1 p_0, \\
p_2 &= \rho_2 p_1 = \rho_1 \rho_2 p_0, \\
&\ \ \vdots \\
p_k &= p_0 \prod_{j=1}^{k} \rho_j, \\
&\ \ \vdots \\
p_n &= p_0 \prod_{j=1}^{n} \rho_j.
\end{aligned} \tag{C.30}$$

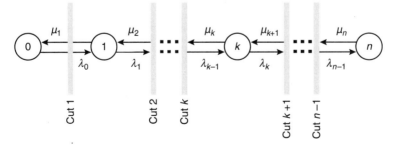

**FIGURE C.4**   "Cuts" of the transition graph for BDP.

Using (C.28), one can write the solution

$$p_0 = \left( 1 + \sum_{k=1}^{n} \prod_{j=1}^{k} \rho_j \right)^{-1}. \tag{C.31}$$

Thus, for any $p_k$, solution is

$$p_k = \frac{\prod_{j=1}^{k} \rho_j}{1 + \sum_{k=1}^{n} \prod_{j=1}^{k} \rho_j}. \tag{C.32}$$

# APPENDIX D

# GENERAL BIBLIOGRAPHY

## HANDBOOKS

Kozlov, B. and I. Ushakov (1966) *Brief Handbook of Reliability Calculations for Electronics Equipment.* Sovetskoe Radio, Moscow (in Russian).

Kozlov, B. and I. Ushakov (1970) *Reliability Handbook.* Holt, Rinehart and Winston, New York.

Kozlov, B. and I. Ushakov (1975) *Handbook of Reliability Calculations for Electronic and Automatic Equipment.* Sovetskoe Radio, Moscow (in Russian).

Kozlov, B. and I. Ushakov (1978) *Reliability Handbook for Electronic and Automatic Systems.* Verlag Technik, Berlin (in German).

Kozlov, B. A. and I. A. Ushakov (1979) *Reliability Handbook for Engineers.* Springer (in German).

Ushakov, I. (ed.) (1985) *Reliability of Technical Systems: Handbook.* Radio i Sviaz (in Russian).

Ushakov, I. (ed.) (1989) *Prirucka Spolehlivosti v Radioelektronice a Automatizacni technice.* SNTL (in Czech).

Rook, P. (1990) *Software Reliability Handbook.* Kluwer.

*Probabilistic Reliability Models*, First Edition. Igor Ushakov.
© 2012 John Wiley & Sons, Inc. Published 2012 by John Wiley & Sons, Inc.

Kececioglu, D. (1991) *Reliability Engineering Handbook.* Prentice Hall.

Ushakov, I. A. (ed.) (1994) *Handbook of Reliability Engineering.* Wiley.

Ireson, W., C. Coombs, and R. Moss (1995) *Handbook of Reliability Engineering and Management.* McGraw-Hill.

Kececioglu, D. (1995) *Maintainability, Availability, and Operational Readiness Engineering Handbook.* Prentice Hall.

Pecht, M. (ed.) (1995) *Product Reliability, Maintainability, and Supportability Handbook.* CRC Press.

Pahm, H. (2003) *Handbook on Reliability Engineering.* Springer.

Misra, K. (2008) *Handbook of Performability Engineering.* Springer.

Stapelberg, R. (2009) *Handbook of Reliability, Availability, Maintainability and Safety in Engineering Design.* Springer.

## TEXTBOOKS AND MONOGRAPHS

Bazovsky, I. (1961) *Reliability Theory and Practice.* Prentice Hall.

Lloyd, D. K. and M. Lipow (1962) *Reliability: Management, Methods, and Mathematics.* Prentice Hall.

Polovko, A. M. (1964) *Fundamentals of Reliability Theory.* Nauka (in Russian).

Barlow, R. E. and F. Proschan (1965) *Mathematical Theory of Reliability.* Wiley.

Gnedenko, B. V., Yu. K. Belyaev, and A. D. Solovyev (1965) *Mathematical Methods in Reliability Theory.* Nauka (in Russian).

Polovko, A. M. (1968) *Fundamentals of Reliability Theory.* Academic Press.

Gnedenko, B. V., Yu. K. Belyaev, and A. D. Solovyev (1969) *Mathematical Methods in Reliability Theory.* Academic Press.

Amstadter, B. L. (1971) *Reliability Mathematics: Fundamentals; Practices; Procedures.* McGraw-Hill.

Barlow, R. E. and F. Proschan (1975) *Statistical Theory of Reliability and Life Testing.* Holt, Rinehart and Winston.

Ryabinin, I. (1976) *Reliability of Engineering Systems. Principles and Analysis.* Mir (in Russian).

Kapur, K. C. and L. R. Lamberson (1977) *Reliability in Engineering Design.* Wiley.

Kaufmann, A., D. Grouchko, and R. Cruon (1977) *Mathematical Models for the Study of the Reliability of Systems.* Academic Press.

Barlow, R. E. and F. Proschan (1981) *Statistical Theory of Reliability and Life Testing*, 2nd ed. To Begin With.

Gnedenko, B. V. (ed.) (1983) *Aspects of Mathematical Theory of Reliability*. Radio i Sviaz, Moscow (in Russian).

Osaki, S.and Y. Hatoyama (eds) (1984) *Stochastic Models in Reliability Theory*. Springer.

Birolini, A. (1985) *On the Use of Stochastic Processes in Modeling Reliability Problems*. Springer.

Melchers, R. E. (1987) *Structural Reliability: Analysis and Prediction*. Wiley.

Rudenko, Yu. N. and I. A. Ushakov (1989) *Reliability of Power Systems*. Nauka, Novosibirsk (in Russian)

Hoyland, A. and M. Rausand (1994) *System Reliability Theory: Models and Statistical Methods*. Wiley.

Gnedenko, B. V. and I. A. Ushakov (1995) *Probabilistic Reliability Engineering*. Wiley.

Barlow, R. E. (1998) *Engineering Reliability*. SIAM.

Aven, T. and U. Jensen (1999) *Stochastic Models in Reliability*. Springer.

Modarres M., M. Kaminsky, and V. Krivtsov (1999) *Engineering and Risk Analysis: A Practical Guide*, 2nd ed. Marcel Dekker.

Bazovsky, I. (2004) *Reliability Theory and Practice*. Dover.

Kolowrocki, K. (2004) *Reliability of Large Systems*. Elsevier.

Neubeck, K. (2004) *Practical Reliability Analysis*. Prentice Hall.

Xie, M., K.-L. Poh, and Y.-S. Dai (2004) *Computing System Reliability: Models and Analysis*. Kluwer.

Zio, E. (2007) *An Introduction to the Basics of Reliability and Risk Analysis*. World Scientific.

Epstein, B. and I. Weissman (2008) *Mathematical Models for Systems Reliability*. CRC Press.

Ushakov, I. A. (2009) *Theory of System Reliability*. Drofa, Moscow (in Russian).

Tobias, P. A. and D. C. Trindade (2010) *Applied Reliability*, 3rd ed. CRC Press.

# INDEX